BIOCHEMISTRY of PHOTOSYNTHESIS

2nd Edition

BIOCHEMISTRY of PHOTOSYNTHESIS

2nd Edition

R. P. F. Gregory
Department of Biochemistry,
The University of Manchester

A Wiley-Interscience Publication

JOHN WILEY & SONS LTD
London · New York · Sydney · Toronto

Library of Congress Cataloging in Publication Data:

Gregory, Richard Paul Fitzgerald, 1938–
   Biochemistry of photosynthesis.

   Includes bibliographies and index.
   1. Photosynthesis. I. Title.
QK882.G83   1976   581.1'3342   76–13501

ISBN 0 471 32676 3

Text set in 11/12 pt. Photon Times, printed by photolithography,
and bound in Great Britain at The Pitman Press, Bath

Foreword

The first edition of this book was a valuable addition to the scientific literature. That a second edition is to appear after an interval of five years concides with what I believe to be a considerable revival of interest in the subject. This seems to be due not only to a wider realization of our dependence on energy received from the sun but also to the advancing knowledge of membranes and their function in cellular activities. It happened that few of the other textbooks were written from so definite a biochemical approach because photosynthesis had long been studied by techniques which belonged to other branches of science. Dr. Gregory succeeds well in showing the impact of these on his subject and has introduced them so that they can be accessible to students of biochemistry. Surely, this new edition will help to foster the wonder we all can have for the marvellously efficient means of conversion of radiant energy developed through the evolution of living cells.

Preface to the first edition

The process of photosynthesis, in which the energy of light (sunlight) is chemically captured by living organisms, and on which the whole of our planet's life-stock depends, is worthy of considerable stress in an undergraduate course of biochemistry, the more so if it provides for a synthesis of topics normally isolated by the necessarily linear nature of such courses. For example, the metabolic pathways of photosynthesis form a useful antithesis in the understanding, at an elementary level, of the pathways of glycolysis and the pentose cycle. The same applies to the cytochromes of the thylakoid, the production of oxygen from water, NADP reduction and photophosphorylation, all of which can be profitably compared with analogous processes in mitochondria. The chloroplast itself, in its relation to the cell, presents a most striking example of biochemical compartmentation. The first aim of this text is to provide an introduction to photosynthesis on the above basis. It is hoped that this introduction, Part I, will be with some selection valuable in courses of botany, and possibly at sixth-form level as well.

Since an introduction involves simplification, which is somewhat unsound, the text continues in Part II towards a second purpose, that of presenting an account of the subject giving the principal points of view (in 1970) and the experimental work and argument by which they were defended. As the writing progressed, however, it became clear that simplification would reappear, and the reader should be aware of two serious manifestations of it. First, the historical aspect and the development of the present-day concepts of photosynthesis, has been largely suppressed, except where old or obsolescent terms and hypotheses may still be unearthed by students and cause confusion. References in the text, and figures reproduced from previous publications, have in the main been selected for their utility and clarity of exposition rather than for evidence of priority. Secondly, the style adopted for the work is that of dividing up the field into a few areas, one to each chapter, and within each chapter to set out sections each written round a discrete point of view, aiming to cover the greater part of the area concerned.

In accord with the aim of supporting a course of lectures in this subject, I have given at the end of Part I a selection of problems, some numerical, some requiring discussion. To enable the most use to be made of Part II as a reference section, the index has been laid out in an extended manner, for which I am grateful to the publishers.

It is a pleasure to thank all my colleagues who have helped with suggestions, particularly Dr. I. West and Dr. A. G. Lowe. I am indebted to Mr. A.

Greenwood, Dr. H. Bronwen Griffiths, Mr. R. Bronchart, Professor R. B. Park and Dr. G. Cohen-Bazire for electron micrographs, and Professor D. A. Walker for an autoradiograph. I gratefully acknowledge the painstaking criticism offered by Professor F. R. Whatley, Professor Walker and Dr. A. R. Crofts. My thanks are also due to Mrs. D. M. Warrior for typing the manuscript, and above all to my wife Julia for constant help and encouragement.

RICHARD GREGORY

Department of Biological Chemistry,
The University of Manchester.
Spring 1971.

Preface to the second edition

It is a privilege to have an opportunity of revising one's work. Apart from correcting errors, which I regret having made and still more my students having seen, it is a pleasure to record the advances made, particularly in the two fields of carbon metabolism and photophosphorylation. The layout and style of the first edition have been preserved. While the text cannot claim to be a review, it is hoped that the student will find some account of most major experiments and discussion up to the end of 1975.

I am most grateful to Professors C. C. Black and A. Staehelin for electron micrographs, to Professors D. O. Hall and D. A. Walker who provided most helpful criticism and gave me access to material in advance of its publication, to Professor V. Massey for Figure 8.1(d), and to Dr. M. C. W. Evans for help with Table 8.2. Dr. J. Barber, Dr. S. Raps and Dr. R. Hill were generous with their time in reading and discussing the manuscripts; but they were in no way responsible for deficiencies remaining. I thank Mrs. J. Black for ably typing the manuscript, and Mr. H. Mann for his help with the task of indexing.

R. P. F. GREGORY
Manchester
February 1976.

Contents

Part 1

Part 7

CHAPTER 1

The context of photosynthesis

1.1 The energy of life

The biochemist investigates biological problems using the techniques of chemistry. Living matter is made up of materials which can be separated and analysed, and the changes that continually take place can be observed and even copied using extracted substances *in vitro*. One prominent goal is to explain the behaviour of living things in terms of identified chemical materials and their reactions. It is hard to think of any region of biology to which biochemists are foreign. Indeed it is through a biochemical approach that many biological topics may be seen to be related. In this text we shall examine photosynthesis, an activity mainly of green plants, and show how the problems raised by the phenomenon have been investigated, and how the importance of the ideas involved extends out of the world of plants and affects our understanding of the fundamental processes of life.

What processes of life would we regard as fundamental? Let us approach this by considering what, in chemical terms, a living organism is. First, we can recognize a discrete enclosed space, with a skin or membrane round it making a boundary. Inside the space is a structured system with solid and liquid phases containing proteins, lipids, carbohydrates, nucleic acids and other substances, all in an aqueous medium which is kept more or less constant in composition, regardless of wide variations in the external environment. As described so far, such a system could not be expected to persist for very long in the face of continuous disruptive processes. These include spontaneous and random hydrolyses of protein and nucleic acid components, and chemical reactions of many materials with oxygen. Also the boundary-structure, which must not be impermeable, tends to lose or gain water and solutes and thus destroy the internal environment of the organism. Again, the structures of the system are easily damaged by mechanical shock, heat and so on. Damage of this kind occurs continually and has to be made good. If the organism is to maintain its structure and size, and still more if it is to grow, it must have a source of energy available to it from the environment. Given this source of energy it can replace materials as fast as they are degraded, and given a source of component material as well, it can grow.

We therefore look for sources of energy available to living organisms. Only two are found, and these represent a clear dichotomy in biology. The first source is chemical energy. This is represented by a substance or mixture of substances that is unstable, meaning that an irreversible process is liable to occur to change the substance. When such a reaction takes place, we say that the system loses

3

energy,* or that energy is released. The organism takes in the unstable material, and causes the reaction to proceed faster. This is catalysis, and the biological catalysts are termed enzymes. They are arranged so that the energy released by the reaction is partly conserved in a form which the organism can apply to the processes of repair and growth (see Figure 1.1). An example of a single substance providing energy by a catalysed breakdown is glucose undergoing fermentation to ethanol and carbon dioxide by yeast. Glucose and oxygen together yield considerably more energy in the chain of reactions leading to carbon dioxide and water. (There is an important chemical principle (Hess' Law) which states that the quantity of energy released by a process depends only on the initial and final

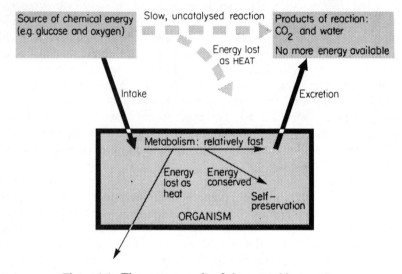

Figure 1.1. The energy supply of chemotrophic organisms

states of the system, not on the nature of the intermediate pathways.) Most of the materials that yield chemical energy for life are in fact produced by living organisms, so that forms of life that require chemical energy (we shall term these *chemotrophs*, and the life-style *chemotrophy*) consume the dead or living bodies of other organisms. It must be clear that as energy is lost to the environment, usually as heat, at each turn of the food cycle, so chemotrophic life by itself cannot last very long before a state of famine arises. Another source of energy is needed.

The second source is the energy of light. Of many other kinds of environmental energy source, thermal, gravitational, seismic, radioactive and electric, only electromagnetic radiation has been of any biochemical value, and of the immense range of such radiation, only that part known as visible light. The principle by which light interacts with matter and enables the organism to abstract its energy

* Usually heat is released. This is not always the case; if the system increases in disorder, it may proceed even if heat is absorbed. We use the term *free energy* to avoid the confusion. Free energy (G) combines heat (H) and disorder units (entropy) (S) in the relationship $\Delta G = \Delta H - T\Delta S$ (at constant temperatures).

is interesting and instructive. Of the range of wavelengths of electromagnetic radiation, shown in Figure 1.2, those of the visible light band result in the formation of *electronically excited states* of certain types of substance (pigments). Longer wavelengths are strongly absorbed by water and very many biological materials, but give rise instead to the excitations detected as heat. Shorter wavelengths, on the other hand, do give rise to electronically excited states, but not only are these states so energetic that they cause random reactions harmful to the organism, but there are relatively many materials which absorb.

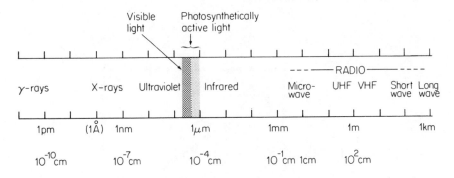

Figure 1.2. The range of biologically useful light in in the spectrum of electromagnetic radiation

Ultraviolet light is of no use to the organism, at least for the purposes of providing energy, and it appears that a screen of special pigments is often present that absorbs the harmful radiation safely.

Organisms that require light energy (termed *phototrophs*) form a dense screen of pigment that absorbs light at a wavelength long enough to avoid screening by other cell materials, and short enough for the energy to be trapped in an electronically excited state. This excited state is unstable; it will lose its energy by one of several processes. Although its lifetime is short, the phototroph acts to shorten it further, by catalysing its decay along a specific path, from which some of the energy can be conserved in a form which the organism can apply to the processes of repair and growth (see Figure 1.3). The similarity between chemotrophy and phototrophy is very marked. The biochemical process of obtaining energy from chemical sources is *respiration*, and that in which light energy is utilized is *photosynthesis*.

We have divided living organisms into the chemotrophs and phototrophs on the basis of the source of their energy. An alternative classification is that based on the nature of the nutrient that provides carbon material for the growth of the organism. *Autotrophs* need only carbon dioxide, while *heterotrophs* require organic substances. Earlier, autotrophic nutrition was considered to be synonymous with photosynthesis in plants, and heterotrophic nutrition with respiration in animals and bacteria. However the discoveries, first of the *chemosynthetic bacteria* which are able to grow on carbon dioxide, obtaining their energy from the oxidation of inorganic materials such as sulphur or ferrous iron,* and secondly the *photosynthetic bacteria* which use light energy to grow on materials which may be either organic or inorganic, made this scheme of classification somewhat cumbersome. Thus the chemosynthetic bacteria were termed 'chemiautotrophic', and the two life-styles of photosynthetic bacteria 'photoheterotrophic' and 'photoautotrophic' respectively. As will be shown later in this text, several photosynthetic bacteria are able to practise either mode, depending on what substrates are available. All things considered, it seems easier and of more fundamental importance to make our distinction on the basis of the

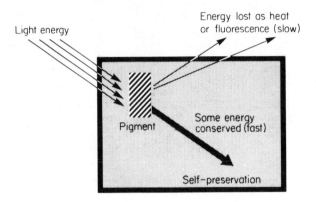

Figure 1.3. The energy supply of phototrophic organisms

energy source rather than on the nature of the carbon supply. Table 1.1 shows how the two classes intersect.

Figure 1.4 represents the passage of energy from light into the chemical form of the materials synthesized by the phototroph, and its subsequent utilization by the chemotroph. Heat is lost at every stage, so that in a steady-state biological situation, all the energy of light is eventually converted to environmental heat. The figure illustrates a further point: the chemical materials which make up the energy source for the chemotroph are in the main also the source of the material from which, with the energy, the chemotroph synthesizes new material and grows. In the case of the phototroph, however, the nutrient material is often the stable waste materials of the chemotroph, such as carbon dioxide and water, from which no more chemical energy is available. In a steady-state system these nutrient cycles must balance. This balance is the basis of the familiar carbon-cycle diagram (see Figure 1.5). While this diagram shows small backwaters of the cycle, mostly involving bacterial fermentations and the heterotrophic photosynthesis, it stresses that the processes of photosynthesis and respiration are closely interdependent. Even the processes carried on by the chemosynthetic bacteria are dependent, since their energy sources, oxygen, nitrates, ferrous iron

---

* See Rabinowitch (1945a).

Table 1.1. Classifications of organisms

		Carbon source	
		CO₂ (autotrophic)	Organic (heterotrophic)
	Chemical (Chemotrophic)	Chemosynthesis (bacteria)	Respiration
Energy source	Light (Phototrophic)	Photosynthesis (green plants and some bacteria) Photoautotrophic	Photosynthesis (bacteria) Photoheterotrophic

and sulphur compounds are mainly produced by other organisms, the chief exception being the rare exposure of mineral sulphur.

The carbon cycle diagram, Figure 1.5, needs an indication of scale. Various estimates have been made of the annual quantity of carbon dioxide fixed by

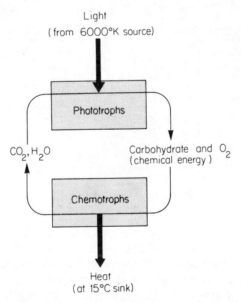

Figure 1.4. The dependence of chemotrophs on phototrophs, and of phototrophs on light from the sun, in terms of a flow of energy

photosynthesis, the main difficulty being that the greater part takes place in the ocean. The ocean indeed contains most of the available carbon dioxide, although the atmosphere is the principal reservoir of oxygen. The mass of the atmosphere, $5 \cdot 3 \times 10^{18}$ kg, includes $1 \cdot 1 \times 10^{18}$ kg oxygen; the sea has been calculated to contain some $5 \cdot 0 \times 10^{16}$ kg carbon dioxide. Taking Riley's (1944) figure for the annual fixation of carbon dioxide as $5 \cdot 7 \times 10^{11}$ tons ($7 \cdot 0 \times 10^{14}$ kg), the carbon dioxide of the 'biosphere' must be turned over every few hundred years. If we

8

assume that the process of photosynthesis averaged out over the earth can be represented by the equation

$$CO_2 + H_2O \longrightarrow O_2 + \tfrac{1}{6} C_6H_{12}O_6$$

we know that for each mole of carbon dioxide fixed a mole of oxygen is released.

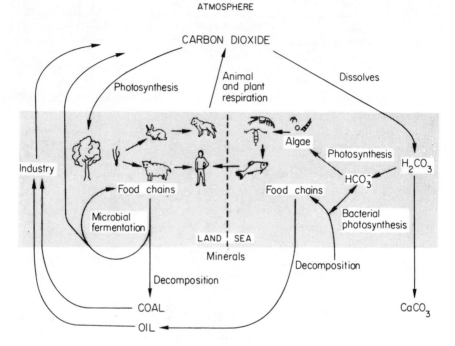

Figure 1.5. The carbon cycle

The annual rate of oxygen release corresponding to the above figures is $5 \cdot 1 \times 10^{14}$ kg,* which must turn the reservoir over in some 2000 years. These time scales are short; if the processes of respiration and photosynthesis were not in close balance, the composition of the atmosphere would have changed significantly even in so short an interval as historical time. We have to conclude that the activities of the chemotrophs and phototrophs are in exact balance.

1.2 The nature of light

The nature of light has been a conceptual problem for many hundreds of years. Newton had regarded light as a stream of particles, a view which was later held to have been invalidated by the demonstration by Young of interference phenomena, rationalization of which was only possible on the basis of a wave

---

* This figure $5 \cdot 1 \times 10^{14}$ kg is obtained by multiplying $7 \cdot 0 \times 10^{14}$ kg, the uptake of CO_2 by 32/44, the ratio of the molecular weights of O_2 and CO_2.

theory, as shown in Figure 1.6. This view persisted, in spite of difficulties about the nature of the medium (the luminiferous ether) required to support the wave motion, until the beginning of the twentieth century. In 1905 Einstein showed that the photoelectric effect presented formidable problems. In the photoelectric effect, light incident on a suitable surface in a vacuum causes electrons to be emitted from it. The energy of the electrons is independent of the intensity of the light, which only affects their number, but is strongly dependent on the wavelength of the light (see Figure 1.7). Regardless of the intensity of the light, emission commenced with no perceptible time lag. It was necessary to conclude

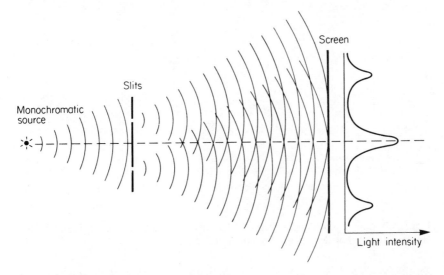

Figure 1.6. The phenomenon of interference in a beam of light passing through a double slit

that all the energy incident on the surface as a whole could more or less instantaneously be gathered at a single atom and result in an electron emission, which made no sense on a simple wave theory. Secondly, the existence of a red limit on the wavelength of light for electron emission agreed with the theorizing of Planck five years previously; Planck had derived an expression for the wavelength distribution of radiation emerging from a cavity in terms of the temperature of the surrounding material. This expression, which agreed with experimental findings, was based on the novel assumption that energy in the atoms of the material was quantized, that is to say was only to take discrete values. In the transition from one energy state to another, a definite *quantum* of radiation energy was emitted, of which the wavelength could be calculated from the equations

$$E = h\nu \qquad \nu = c/\lambda$$

where E is the energy, ν the frequency, h a universal constant called Planck's constant, λ the wavelength and c the speed of light. In 1905 the photoelectric effect was explained in terms of quanta, particles or wave-packets, small enough to interact with atoms or even electrons, with energies governed by their wavelength. A certain energy was necessary to detach the electron, and additional energy increased the kinetic energy of the expelled electron, thus explaining Figure 1.7.

This view was further modified. It was found that beams of electrons displayed interference and diffraction behaviour. An electron appeared to possess a characteristic 'wavelength' depending on its energy. Yet electrons were thought to be material objects. This surprising result was rationalized by Heisenberg in

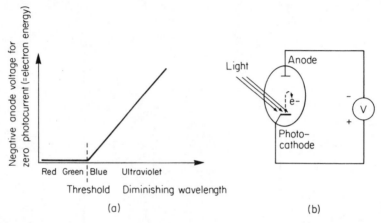

Figure 1.7. The photoelectric effect: (a) The existence of a threshold wavelength of light necessary to release electrons in the apparatus (b).

the *Uncertainty Principle*, which sets a limit on the accuracy with which the position and momentum of a particle can simultaneously be known. This limit is so small that for particles as large as atomic nuclei, and of course all familiar material objects, it has no significance. However for light particles such as electrons and *photons* (a photon being the particle carrying a quantum of radiation) the area of uncertainty in the position of a particle of defined momentum is so great that it has to be expressed in terms of statistical mechanics, or wave-mechanics, which yields a pattern. This pattern has the form of a wave, but the only interpretation of the wave-pattern is that the square of the amplitude at a given point is proportional to the probability of finding the electron there. If by means of an observation the position is actually determined, the wave pattern vanishes, and if the particle continues to move from that point onwards, it has an unknown momentum, and a new probability pattern is formed to describe its future appearance. Thus an electron or a photon allowed to fall on a surface with two slits as in Figure 1.6 has a probability of being found at points on the second surface given by the wave-pattern we know as interference fringes. In between the generation of the particle and its final detection, its position is not determined, and any attempt to find its path interferes with its momentum and ruins the experiment. This is apparently a fundamental uncertainty: there is no possibility of finding, for example, which slit it passed through in Figure 1.6. The usual attitude is to say that what can never be verified does not exist, or at least is of no concern to science. One could use the less exact phrase that electrons and photons propagate as waves but are detected as particles.

1.3 Photosynthetic structures

From the point of view of a chemist a living organism such as a spinach plant presents material that is not only heterogeneous but which also has an organization like a nest of chinese boxes; in pursuing the site of the photosynthetic process he will separate the leaves, then the mesophyll cells, then chloroplasts, and then proceed to disaggregate the chloroplast components. The further this is taken the greater the suspicion aroused in the mind of the biologist, who insists that the relations between parts are as important as the parts themselves. The fact remains however that in green plants the biochemical part of photosynthesis is almost entirely located inside the chloroplasts, which are convenient particles to prepare and investigate. Without implying that it is independent of its environment, we shall find the chloroplast and its components a useful system on which to base a biochemical survey of photosynthesis. Furthermore, although the structures and processes differ in the photosynthetic bacteria and the blue-green algae, which do not have chloroplasts, we shall find that it is convenient to study them in comparison with green plants rather than as an independent subject.

Plate 1 (p. 12) shows a section through a mesophyll cell of a broad bean leaf. The chloroplasts show up clearly as heavily-staining objects, which account for a considerable part of the protoplasmic mass. The layer of cytoplasm that covers the nucleus, chloroplasts and cell wall is very thin; most of the volume enclosed by the cell wall is the vacuole, a watery solution which may act as a waste-pool or a reservoir of water or solutes. This is typical of photosynthetic cells of higher plants. It is not surprising that the chloroplasts may contain up to 70% of the total cell protein. Chloroplasts vary in size and shape. Those of spinach tend to be some 5–7 μm (1 μm $= 10^{-6}$ m), and oblately spheroidal in shape. They tend to be distended by starch grains, or in algae by pyrenoids, which are temporary stores of starch and protein respectively.

The chloroplast illustrated in Plate 2 (p. 13) shows an external double membrane, each layer of which is a 'unit' membrane of the type found in many cell membranes such as the endoplasmic reticulum. This double membrane is termed the chloroplast envelope. (The term 'chloroplast membrane' can be confused with the lamellar system inside.) The envelope is osmotically sensitive: if a leaf is cut with a razor in a drop of isotonic sucrose on a microscope slide, a few intact chloroplasts are released and may be observed under the microscope. Addition of a drop of water causes characteristic balloons to appear. Sometimes, when sections are observed under the electron microscope, 'blisters' are observed between the unit membranes forming the envelope. When chloroplasts are isolated from leaves by grinding and centrifugation, the envelope is easily damaged, and up to 1963–1965 this almost inevitably happened when the

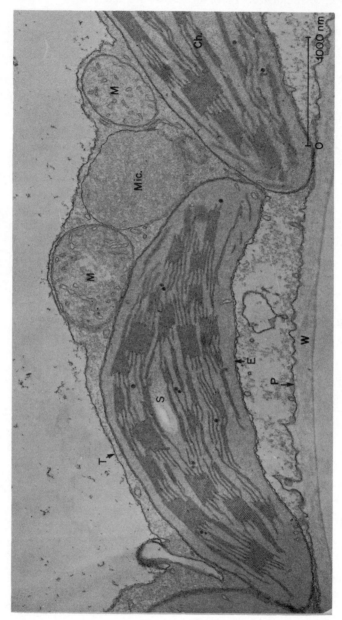

Plate 1. Shows small part of the cytoplasm of a broad bean mesophyll cell with the main organelles. Chloroplasts can be seen in contact with mitochondria and a microbody (possibly concerned in the glycollic acid metabolism of photosynthesis). The section is cut perpendicular to the membranes of the thylakoids showing that these lie predominantly parallel to the cell wall. Ch—Chloroplast. E—Envelope. M—Mitochondrion. Mic.—'Microbody' (glyoxisome? peroxisome?). P—Plasmalemma (cell outer membrane). S—Starch. T—Tonoplast membrane. Vac.—Vacuole. W—Wall (of cell). Courtesy of A. D. Greenwood, Botany Department, Imperial College, London.

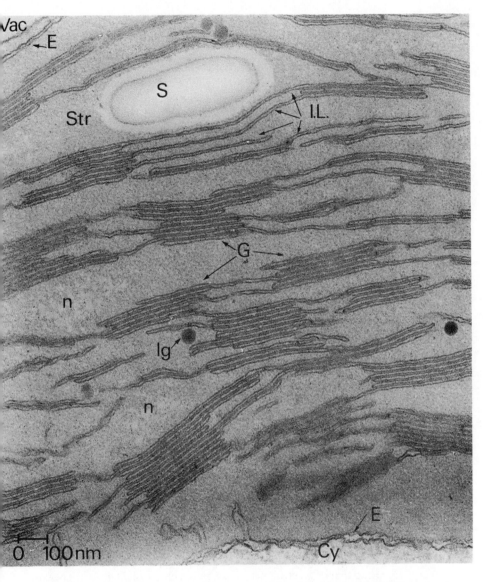

Plate 2. Part of a broad bean chloroplast to show the construction of the grana and the intergranal lamellae. E—Envelope. I.L.—Intergranal lamellae (intergranal thylakoids). G—Granum. lg—Lipid globule (plastoglobulus, osmiophilic globule). n—Nucleoid (DNA-containing region). S—Starch grain (note inner and outer regions). Str—stroma. Vac—vacuole of cell. Cy—cytoplasm of cell. Courtesy of A. D. Greenwood, Botany Department, Imperial College, London.

chloroplasts were washed free of cytoplasm. Damage causes loss of the *stroma*, which is the ground-substance staining moderately with electron-dense stains and which fills the chloroplasts and surrounds the lamellar structure. Such chloroplasts retain their general form, but there is some increase in size, and under the light microscope they lose their bright, refractive appearance and seem 'flat'; in most higher plants *grana* (spots of darker green) can be seen after this treatment.

The most striking feature of electron micrographs of chloroplasts is the system of lamellae. On inspection these turn out to be closely-flattened sacs termed thylakoids ('bag-objects'). The extent of the thylakoid system in any chloroplast varies both with the species and with the history of the individual plant. In many green plants including spinach there are many regions where the thylakoids are collected into stacks; these stacks are the *grana* referred to above. Individual thylakoids may continue from one granum to another, but the thylakoids in a granum (often known as discs, the granum being a stack of discs) appear to stain more densely or are thicker-walled than those outside the granum. Disruption of chloroplasts in a homogenizer or by ultrasonic treatment allows the thylakoid fragments to be collected, free from stroma and envelope. All the pigments of the photosynthetic system, chlorophyll and carotenoids, are present in the thylakoid membranes. The chloroplast is thus divided into three regions: the stroma, the 'solid' material of the thylakoid membranes, and the very small space inside the thylakoids. This separation of phases is important.

Also visible in the chloroplast are 'osmiophilic globules' (also termed plastoglobuli), so called because they take up the electron-dense stain osmium tetroxide, which contain lipid material. With special staining techniques, deoxyribonucleic acid (DNA) can be seen in the stroma. (DNA is principally found in cell nuclei, where it is part of the mechanism of Mendelian inheritance; it has been shown that the sequence of nucleotides that make up DNA governs the formation of at least some of the cell proteins.) The presence of DNA in the chloroplast raises interesting problems concerning the degree of independence of chloroplasts from nuclear genetic control. Under special conditions, ribosomes, the seat of protein synthesis in the cell, can be seen in the chloroplast as well. Both the DNA and the ribosomes of chloroplasts can be distinguished biochemically, in homogenates, from bulk of the DNA and ribosomes in the cell.

The blue-green algae (Cyanophyta) carry out an autotrophic photosynthetic process similar to that of green plants; they contain chlorophyll *a* as the principal pigment, and evolve oxygen. However there are some important differences: the thylakoids are not contained in chloroplasts, but appear to ramify throughout the cell (Plate 3 p. 16). The cytoplasm of the cell represents the stroma of the chloroplast of the higher plant. The algae show differences in their photosynthetic pigments: in green algae and higher plants chlorophyll *b* is an *accessory pigment* to chlorophyll *a*. In some other groups of algae chlorophyll *c* appears instead of chlorophyll *b*, and in others, including the blue-greens, there is no accessory chlorophyll but instead various *biliproteins*, which are formed by the combination of protein with various *bile pigments*, which while derived chemically from compounds of the chlorophyll type, are by no means similar in properties.

The photosynthetic bacteria present a very different picture, both in their organization and in the biochemical details of their photosynthetic process. They have been divided into the *purple* and the *green* bacteria, and the purple group has been divided further into *sulphur* and *non-sulphur* purple bacteria. However the distinctions between these groups are by no means clear cut. In general these bacteria have cell walls similar to those of other gram-negative bacteria, inside which is a thick cell membrane. This cell membrane often appears somewhat folded, and may give rise to membranes or vesicles within the body of the cell. When the cells are disrupted, these membranes may be obtained by centrifugation, and are termed *chromatophores* (not to be confused with chemical *chromophores*). The membrane system *in vivo* may or may not be recognizable as discrete chromatophores. Plates 4 to 7 (pp. 17–20) illustrate the membrane material in four species of photosynthetic bacteria.

The membranes of purple bacteria contain not chlorophyll *a* but bacteriochlorophyll, which is closely related chemically to chlorophyll *a* but absorbs further into the red; *in vivo*, the pigments can make use of light well into the infrared region. The colour of the purple bacteria is due mainly to carotenoid pigments. Carotenoids (including the carotene and xanthophyll compounds) are universal in photosynthetic organisms, but they are not all necessarily involved in the capture of light for energy conversion in photosynthesis. Another role appears to be a protection in some way from the combined effects of oxygen and light which damage living matter.

While in the green plants carbon dioxide is reduced using hydrogen from water (leaving oxygen to be evolved as a gas), in the photosynthetic bacteria this never occurs. Instead other hydrogen donors are employed; in the green and the purple sulphur bacteria sulphide or thiosulphate is used, and in the purple non-sulphur bacteria a variety of organic compounds may be oxidized and the hydrogen used for reduction of the carbon source. The carbon source may be carbon dioxide (autotrophic nutrition) or an organic compound (heterotrophic nutrition) which may even be the same material as was used for the hydrogen donor. It is remarkable that in at least one group, the purple non-sulphur bacteria, metabolism can switch to an aerobic, chemotrophic growth, and the formation of pigments stops until such time as photosynthesis is required again. For the characteristic features of photosynthesis, we must therefore look first not at the metabolic reactions of the hydrogen donor and carbon substrates, but rather at the unique energy conversion and conservation system.

The environmental problem

There is growing anxiety about the ability of the Earth to provide food for all the human population at the level of nutrition desired, and to provide energy (traditionally in the form of 'fossil fuels', coal and oil) for the existing industrial nations and those that intend to industrialize. The first problem has in the past solved itself by the tendency of the most fecund communities to live at the point

16

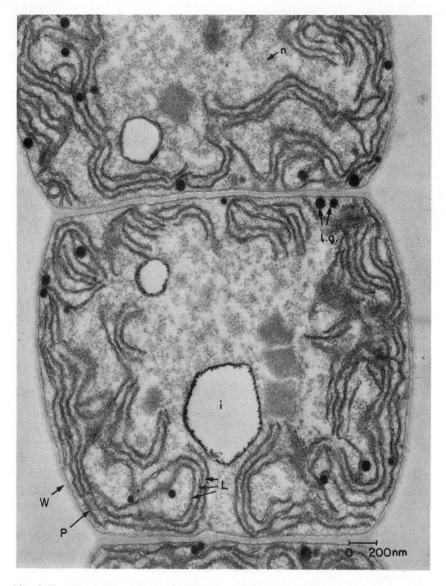

Plate 3. Electron micrograph from a thin section of a blue green alga (*Nostoc sp.*). The chlorophyll-carrying lamellae (L) or thylakoids, are situated in the general cytoplasm of the cell, mainly towards its periphery where lipid globules are also common. As in the red algae each thylakoid is separate from its neighbours with the biliprotein accessory pigments closely associated with the interior of the membranes and aggregated into phycobilisomes but these are often small in size and, as here, not easily distinguished from other cytoplasmic particles. The DNA is found as fine fibrils in the 'nucleoid' (n) regions of the cytoplasm. Other characteristic cell components are polyhedral bodies (P), probably protein, and vesicles (i) which may be polyphosphate. (W)—Wall of cell. Courtesy of Dr. H. Bronwen Griffiths, Botany Department, Imperial College, London.

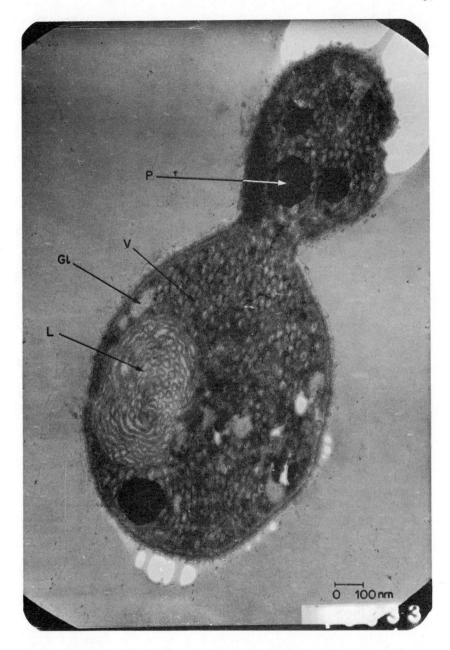

Plate 4. Electron micrograph of a section through the photosynthetic bacterium *Thiocapsa* (Thiorhodaceae). The vesicles (v) and the lamellae (L) may be regarded as corresponding to the thylakoid material of chloroplasts. (P) polymetaphosphate deposit. (Gl) glycogen deposit. Courtesy of Dr. G. Cohen-Bazire, Department of Bacteriology and Immunology, University of California, Berkeley.

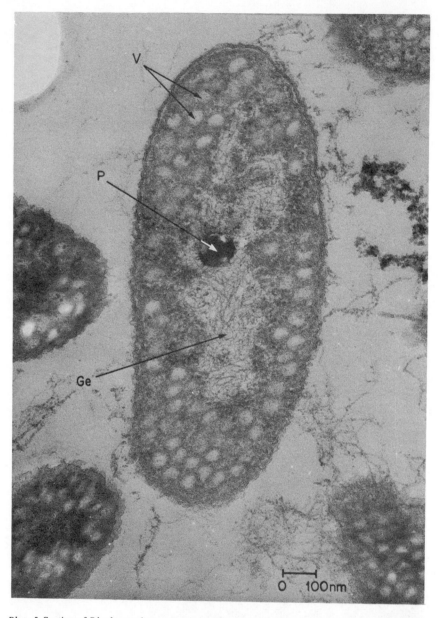

Plate 5. Section of *Rhodopseudomonas spheroides* (Athiorhodaceae) grown photosynthetically at low light intensity. The photochemical apparatus is in the form of 50 nm vesicles (V). At higher light intensities lamellar material is also formed. (P) polymetaphosphate deposit. (Ge) region containing genetic material (DNA). Stained with lead hydroxide. Electron micrograph by courtesy of Dr. G. Cohen-Bazire, Department of Bacteriology and Immunology, University of California, Berkeley.

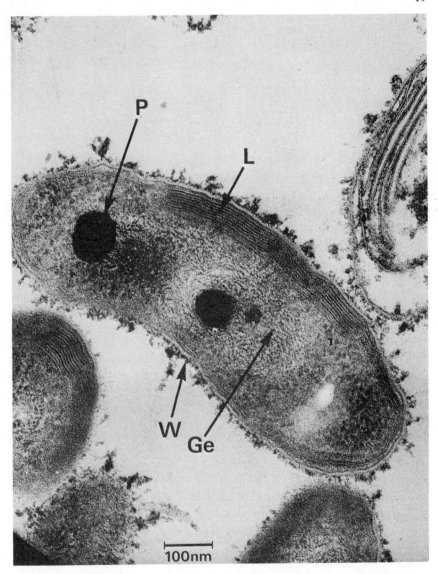

Plate 6. Rhodopseudomonas palustris (compare with Plate 5). Here lamellar material is evident (L). (W) cell wall. (Ge)—genetic material. (P)—polyphosphate deposit. Electronmicrograph by courtesy of Dr. G. Cohen-Bazire, Department of Bacteriology and Immunology, University of California, Berkeley.

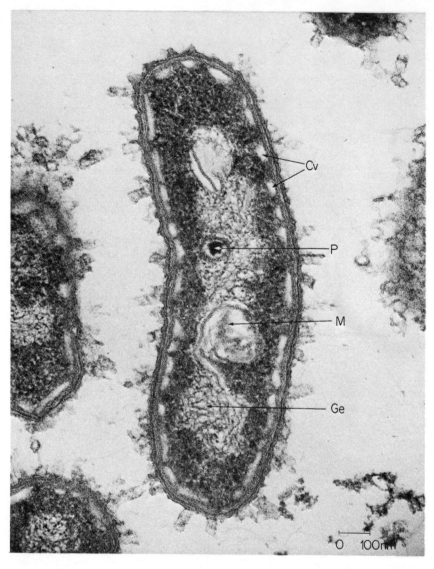

Plate 7. Chlorobium thiosulfatophilum (Chlorobacteriaceae). The vesicles (Cv) *Chlorobium* vesicles) are distributed round the periphery of the cell. (M) mesosome, an enigmatical structure, common among bacteria, which communicates with the exterior and may be involved in DNA transfer. (Ge)—genetic material. (P)—polyphosphate deposit. Micrograph by courtesy of Dr. G. Cohen-Bazire, Department of Bacteriology and Immunology, University of California, Berkeley.

of starvation, and to that extent the increased world population reflects the advance in agricultural activity. However, if an acceptable population limit can be found, the starvation problem can be brought to an end by the application of proven agricultural improvements, including selection of strains and improvement of crops, in relation to specific environments. Thus the prospect of genetically introducing 'C_4' characteristics (see p. 187) and nitrogen-fixation systems into varieties limited by photorespiration or localities limited in nitrogen status is attractive, if difficult.

The food supply at the present time is the problem of 'the third world'. The energy supply is the problem of the industrialized nations. On paper, the energy from the sun that falls on any temperate or tropical country greatly exceeds the energy extracted there from fossil fuels, so that should these become insufficient there is a replacement, at least in theory. As an alternative to mechanical devices for capturing solar heat there is the possibility of conducting industrialized photosynthesis, using efficient plants, algae or bacteria, leading perhaps to the production and distribution of the 'perfect' fuel, molecular hydrogen. Progress in these directions is difficult, and assessment of costs and benefits almost impossible. But the basic knowledge will have to be acquired now, since by the time it is needed it will be too late to start.

A more hopeful prospect is the use of photosynthesis to deal with the increasing problem of waste caused by the populations of man and intensively-farmed animals in industrialized countries. In California, Oswald (1973) has pioneered the use of algae, which now recycle organic waste from towns of moderate size, at a profit. This appears to be only a marginal possibility at latitudes as high as the British Isles, but worldwide it must be safe to assume that we shall see much more use made of this process.

The point should not be lost that attention to photosynthesis is not only of great interest in itself, but is also vitally necessary—the need being none the less desperate for its apparent lack of publicity.

CHAPTER 2

The absorption of light

2.1 Pigments and chromophores

A pigment is a chemical substance that absorbs visible light. That part of the molecule which is responsible for the absorption is termed the *chromophore*. In small molecules the chromophore may be inseparable from the whole structure, just as all the atoms of say formic acid contribute to the observed acidic properties. In larger molecules however certain groups can be recognized as chromophores regardless of the rest of the molecule, just as in general the group ·CO·OH can be recognized as a structure conferring acidic properties. Obviously this is only a matter of convenience, since the exact properties of chromophores (and the dissociation constants of acids) are bound to be affected by the other atoms of the molecule.

In biological materials there are relatively few types of molecular structure that absorb visible light. In section 1.1 it was pointed out that absorption in this region of the spectrum was associated with the excitation of specific electrons of the molecule to excited states; excited states readily underwent specific chemical reactions providing the basis of photosynthesis and vision. In the infrared region the energy of the quanta absorbed by biological systems (which are aqueous) is directly converted to thermal energy. Some bacteria can utilize light of wavelength up to 1000 nm or more for photosynthesis, which indicates rather that the upper limit of visible light, or the lower limit of the energy of electronically excited states, is not precisely defined, than any exception to the rule above. In the ultraviolet region, with the increasing energies of the promoted electrons, the rate of reaction of the activated molecules is likely to be substantial even without catalysis. Furthermore, the number of possible electron transitions in a molecule increases with decreasing wavelength. Hence not only is absorption of ultra-violet light more widespread in biological materials, but the absorption tends to be associated with diverse and uncontrolled reactions.

2.2 Excited states

We will now turn to examination of the process of light absorption in more detail. It must be borne in mind however that the quantitative aspects of this account can only be derived for the simplest molecules, and the pigments that are important in light absorption for photosynthesis are not at all simple. More rigorous accounts of the processes involved can be found in Thomas (1965) or in Seliger and McElroy (1965).

A molecule contains many electrons. Some are 'inner' electrons, such as the $1s$

electrons of carbon, and for our purposes can be ignored. Other 'inner' electrons are the d-electrons of transition elements, and these will be discussed later as a special case. Of the 'outer', or valence electrons, some will be 'non-bonding' ('n-electrons'): they will be centred around one atom, and their behaviour will be only slightly perturbed by the presence of the other centres. Electrons which form bonds are usually electrically centred around two atomic nuclei. If the electron distribution is cylindrically symmetrical around the axis of the bond, it is said to be a sigma (σ) bond; if not, it is said to be a pi (π) bond. In the 'double bond' of, for example, ethylene ($CH_2{=}CH_2$) one bond is π, the other σ. Diagrams are given to illustrate these bond types (see Figure 2.1) but it should be stressed that they do not illustrate the situation in any particular case. Electrons are restricted to orbitals in bonding just as in non-bonding cases. Each orbital

Figure 2.1. To illustrate the symmetries of electron patterns constituting σ- and π-bonds.

type has its family of excited states. In thermal qualibirium at room temperature, a population of molecules has a large majority of electrons in the lowest, or 'ground' states of their respective orbitals. The excited states can be written (σ^*), (π^*) for (σ), (π) orbitals respectively. We can therefore write out a variety of types of electronic transition: $\sigma \to \sigma^*, \pi \to \pi^*, n \to \sigma^*, n \to \pi^*$. It happens, however, that the energy gap between the ground level and the first excited state of virtually all σ-bond electrons is so great (of the order of 10 eV) that the wavelength of the related photon lies in the far ultraviolet (124 nm for 10 eV). So far as biological systems are concerned, we may confine our attention to the $\pi \to \pi^*$ and $n \to \pi^*$ transitions, which may have energies corresponding to the range of visible light.

Figure 2.2 relates the two principal absorption bands of chlorophyll to the formation of the 'first' and 'second' excited states of the molecule. Although the energy stored by the second excited state is much greater than that in the first, the excess is not available for photosynthesis: the second excited state loses its energy in a transition to the first excited state, with the evolution of heat, in an extremely short time. Thus the result of the absorption of a 'blue' quantum is the same excited state as would have been produced by a 'red' quantum directly.

The figure also indicates the manner in which the ground state and the excited states each possess a series of *vibrational substates*. At biological temperatures there are several substates populated at each level, so that there is a range of electronic transitions with slightly different energies. This is one of the reasons for the observed absorption bands being broad; at liquid nitrogen temperatures (77°K) spectra tend to be sharper. Another broadening influence is the tendency of

24

chlorophyll molecules to aggregate in solution, and even more so in the chloroplast lamella. The absorption bands are broader and displaced to the red, and represent transitions of the aggregate rather than of individual chlorophyll molecules.

It may be appropriate to mention here the transitions of the d-electrons of transition metals. In the atoms of these elements d–d transitions are 'forbidden', that is to say take place relatively rarely, so that the light absorption is not very

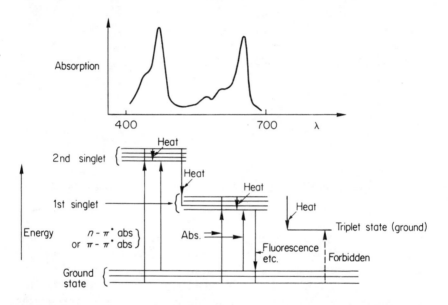

Figure 2.2. The relationship of the principal absorption peaks of chlorophyll with the excited states of the molecule. (The vibrational substates are represented impressionistically)

intense. However in complexes of the metals it may happen that an electron may pass from an orbital of one atom to one on another. This is termed a charge-transfer absorption, and is intense. A good example is the intense absorption of red light operating the transition of an electron between ferro–ferric iron atom pairs in Prussian blue (ferrous ferricyanide). In the chloroplast there is a copper protein, plastocyanin, which also shows an intense absorption (relative to say copper sulphate, atom for atom of copper) which may be due to such a transition. Charge-transfer excitations of transition-metal elements are not, however, likely to play much part in photosynthesis, because the absorption by other pigments is many orders of magnitude greater in the chloroplast. Apart from these d–d transitions, charge transfer complexes also occur in organic systems and may have a secondary importance in photosynthesis. For the present it would appear that $n \rightarrow \pi^*$ or $\pi \rightarrow \pi^*$ transitions of chlorophyll are the basis of light absorption in the chloroplast.

2.21 The triplet state

An orbital can contain up to two electrons, which must then have 'anti-parallel' spins. This may be visualized as in Figure 2.3 where the arrows represent the magnetic axis resulting from the rotation of the electronic charge.

The two directions are arbitrarily given *spin quantum numbers* $+\frac{1}{2}$ and $-\frac{1}{2}$. When one of the electrons is promoted to a higher orbital, by absorption of a quantum of radiation or by other means, the direction of spin of the electron is not normally changed. The algebraic sum of the spin quantum numbers of all the electrons of the molecule, known as the *resultant spin, S,* is unchanged; in the case of chlorophyll in which all electrons are paired, $S = 0$. However there is a

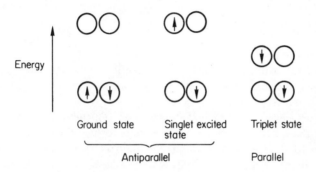

Figure 2.3. The formation of the triplet state preserves the resultant spin value, but reversal takes place when the triplet is formed

small probability that spin reversal may take place during the lifetime of the excited state (see Figure 2.3). When this happens, S becomes 1 (arbitrarily considered positive). This state once formed has a similarly low probability of spin re-reversal so that the lifetime of this new state is longer, and can often be recognized by spectroscopists. The spectroscopic terms for the two types of excited state are 'singlet' and 'triplet' respectively, referring to the *spin multiplicity* $(2S + 1)$. The triplet state may be thought of as the ground state of a new family of states: the first excited triplet, the second, and so on. This is the basis of the spectroscopic observation; the triplet is a new molecule and has a unique spectrum. Chemically, the triplet is more reactive, acting as a *bi-radical*, that is, as it has an unpaired electron in each of two orbitals, it is a free radical twice over. The triplet state is believed to be responsible for most photochemical reactions of dyes in solution, and a hypothesis by Franck and Rosenberg suggested triplet-chlorophyll to be a key intermediate in photosynthesis (see section 8.24). It must be stressed however that because spin reversal is an improbable process, the concentration of triplets remains low even in strong light; in the case of chlorophyll the triplet is hard to detect even in solution, and although it has been detected *in vivo* its relevance is dubious (see p. 49).

2.22 The conversion of the energy of excitation

Fluorescence

The simplest degradation of the first excited singlet state is the reversal of the absorption. The electron falls back to the ground state and a photon is emitted, a process known as fluorescence. In a molecule some of the energy is lost as heat because the vibrations and rotations of thermal energy broaden the energy gap; on absorption of light an electron is in fact promoted to some vibrationally excited sub-state of the electronically excited state. The vibrational excitation is usually lost in 10^{-12} s, while fluorescence takes some 10^{-9} s. Light emitted by fluorescence from a molecule at room temperature is somewhat redder than that which excited it, the difference representing the lost vibrational energy. This is known as the Stokes' shift.

Internal conversion

The excited state of a molecule is not necessarily stable, and the energy may be degraded to thermal energy in vibrational, rotational and (ultimately) translational modes.

Both fluorescence and internal conversions typically take place in about 10^{-9} s. On the other hand if the triplet state is formed the lifetime is extended typically to 10^{-5} s. (Some compounds are known in which triplet states persist for several minutes.) Light emission from the triplet state, which has to be accompanied by spin reversal, is known as phosphorescence and has a longer wavelength than the normal fluorescence for the particular system. Alternatively the triplet can be converted back to the singlet, a process with a low probability. Nevertheless, this requires only a small amount of energy and can be achieved by thermal excitation, and 'delayed' fluorescence with the normal fluorescence spectrum can result.

Collisional deactivation

The activation energy is released as heat by means of a collision with a second molecule. In aqueous solutions at normal temperatures the collision frequency might be of the order of 10^{11} s^{-1}. In the thylakoid membrane, with chlorophyll molecules in a semi-solid state, a lower frequency of collisional deactivation might be assumed.

Energy transfer or migration

An excited molecule can return without collision to the ground state while a second molecule is raised to the excited state. When the molecules are a considerable distance apart (of the order of the wavelength of the light) the mechanism for this process is likely to be fluorescence followed by reabsorption, an inefficient process. In more concentrated solution there is a mechanism, known as *inductive resonance*, in which there is no intermediate photon, the receiver responding to the electric field of the transmitter. This process is far more efficient.

Förster (1959) obtained the following equation for the relation between the probability of a transfer of energy from molecule S to A ($n_{S^* \to A^*}$) and the distance between them (R):

$$n_{S^* \to A^*} = \frac{1}{\tau_s} \left(\frac{R_0}{R} \right)^6$$

where τ_s is the lifetime of excitation in S, and R_0 is a standard distance such that the probability of transfer is equal to that of other deactivation mechanisms. Since

$$\frac{1}{R^3} \propto c$$

where c is the concentration,

$$n_{S^* \to A^*} \propto c^2$$

Csatorday and coworkers (1975) have found that transfer of excitation energy between chlorophyll molecules dissolved in detergent micelles follows the above relationship, suggesting that Förster's analysis (which is based on dipole–dipole interactions) is valid in these systems. Concentrations of 0·1 M (approximately that in the solid phase of the thylakoid) were reached, lending support to inductive resonance as the means of energy-transfer *in vivo*. The topic of energy migration through a solid is interesting and is discussed on its own in section 2.4.

Chemical change

The third means by which an excited molecule can lose its energy is by taking part in a chemical reaction. The pigment molecule may itself undergo change—we are familiar with dyed fibres bleaching in sunlight—or it may not; so far as we can see, the latter is the case in photosynthesis. Again, the light energy may provide only the 'activation energy' required to bring the two reactants together, or, as in the case of photosynthesis, be partly conserved in the chemical energy of the products. Several types of reaction are possible.

Dissociation:

$$A\!-\!B \overset{h\nu}{\longrightarrow} A + B$$

The energy of visible light quanta is in the region of 250 kJ per einstein. (The einstein is 6×10^{23} (Avogadro's number) photons, i.e. a 'quantum mole'.) This does not normally permit rupture of a covalent bond. One example however is the dissociation of the ferricyanide ion by blue light.

Ionization: The excited molecules may lose an electron:

$$AB \overset{h\nu}{\longrightarrow} AB^+ + e^-$$

Since the electron must be accepted by another molecule, this is an oxidation–reduction process. The molecule may alternatively dissociate into positive and negative ions:

$$A—B \xrightarrow{h\nu} A^+ + B^-$$

Thirdly, in a solid state matrix such as a crystal or possibly in the thylakoid membrane, an electron may be expelled which can migrate through the matrix with an extended lifetime. This is the basis of photoconductivity, and will be discussed later.

Association: The excited molecule can combine or otherwise react with a second (unexcited) molecule:

$$A^* + A \rightarrow A_2$$

In a fourth mechanism, *photosensitization*, the excited molecule can sensitize a reaction without appearing to undergo even temporary chemical change. A typical case is the dissociation of hydrogen molecules into atoms, sensitized by mercury vapour:

$$Hg \xrightarrow[253 \cdot 7 \, nm]{h\nu} Hg^*$$
$$H_2 + Hg^* \rightarrow 2H + Hg$$

The energy of the excited mercury (472 kJ mole^{-1}) exceeds that of the dissociation of H_2 by 38 kJ mole^{-1} (see Glasstone and Lewis, 1960).

2.23 Observation of fluorescence in vivo

There is a very striking difference between the strong red fluorescence observed in a dilute solution of chlorophyll *a* in an organic solvent such as acetone or diethyl ether, and the apparent absence of it in a green leaf or chloroplast preparation. In fact fluorescence is emitted from the latter systems, but in a very low yield (of the order of 1%). By suitably protecting the detector (a photomultiplier) from the transmitted and scattered exciting light beam, with filters and a second monochromator, both the emission spectrum and the intensity can be measured. It is found that at room temperature there is a maximum emission at 685 nm; allowing 15 nm for Stokes' shift (loss of vibrational energy on fluorescence) we identify the emission with that part of the chlorophyll *a* which absorbs at 670 nm (see p. 120). The intensity of the emission varies with time, increasing to perhaps twice its initial value in a few seconds (depending greatly on the experimental conditions). This is the Kautsky effect, termed fluorescence induction. It is important (see Chapters 7, 8 and 9). At low temperatures (liquid nitrogen, 77°K) there are three maxima: 685, 695 and 730 nm. These essentially reflect partitioning of chlorophyll among several functionally distinct assemblies. On extinguishing the exciting light, emission continues, at a greatly reduced level, and declines: both magnitude and rate of change being dependent on conditions. This *delayed light emission* is considered to be a back reaction between products of the primary photosynthetic process. With sufficiently sensitive equipment, the emission can be detected for many minutes.

2.3 The time scale

Kamen (1963) arranged the events of photosynthesis diagrammatically so as to emphasize the time scale on which the absorption of light and successive phenomena took place. He used the symbol pt_s, by analogy with pH, for the negative logarithm of the time in seconds. Thus the time taken to promote an electron being 10^{-15} s, the promotion is said to occur at pt_s 15. Figure 2.4 sets out events on such a logarithmic timescale. This approach, which may be unfamiliar, provides a new perspective on photosynthesis, and illustrates in a novel way the contributions made to the subject by various scientific disciplines.

2.4 The solid state: migration of energy and photoconductivity

In section 2.22 there was described the process of inductive resonance by which excitation could be transferred from one pigment molecule to another, in solution. It was shown that a reasonable account of energy transfer could be obtained treating the thylakoid as a solution of chlorophyll. Recently there has grown up a theory of energy migration which applies to crystals, known as exciton theory. Essentially the theory treats the excitation as delocalized; since all points inside a perfect lattice are identical, transfer is considered statistically. An exciton is an energy-packet corresponding to the excitation energy of a unit of the crystal, which there is a probability of finding at any point, and this probability varies according to a wave-equation. Although this theory has been extensively developed for perfect crystals in mathematical terms, it has so far been of little practical value in photosynthesis, since the arrangement of chlorophyll is not that of even an imperfect crystal, although there may be a considerable degree of orientation of chlorophyll. There are however two points that should concern us. The first is that a pigment-array has absorption bands which are different from those of the free molecules.

The second point concerns a form of exciton in which the mechanism is appreciably different from the inductive resonance above. This 'slow exciton' is essentially an electron conduction system. In an electrical conductor there are no energy barriers preventing the outer electrons from moving from atom to atom. In an insulator these barriers are present, and are only broken down by input of energy in destructive amounts. In semiconductors there are effective barriers between electrons in their ground states, but not in their excited states. In a semiconductor an excited electron may wander among neighbouring centres leaving a positive 'hole' (to which it is still electrostatically attracted). The existence of a hole removes the barrier to migration of ground state electrons so that the hole can be filled, creating a different hole; thus both the electron and the hole can migrate through the array, in a sense resembling a free atom. A trap for such an exciton is a centre where either the electron or the hole finds an impurity

30

in the lattice where it is fixed. The fact that films of chlorophyll show photoconductivity (also to a lesser extent preparations of dried thylakoids and bacterial chromatophores) indicates that mobile charges are indeed formed by absorption of light, and an additional attraction of the model is that the reaction centre(s) of photosynthesis that initiate oxidation-reduction reactions have an obvious identity with the electron or hole traps described. Nevertheless the analysis does not

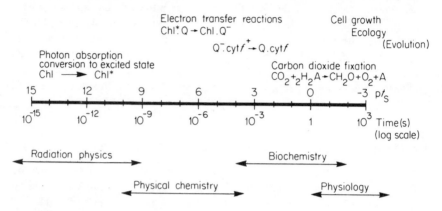

Figure 2.4. The 'pt_s' scale for describing biological activity, after Kamen (1963). The pt_s value is the negative logarithm of the time taken by a given process. From Kamen (1963), *Primary Processes in Photosynthesis*, Academic Press, New York, p. 4, with permission. Copyright held by Academic Press

go much further, mainly because of the lack of known organization in the chlorophyll mass.

2.5 The pigments of photosynthetic systems

Of the various materials that absorb visible light and are found in photosynthetic structures, not all are to be regarded as photochemical sensitizing agents. Thus the protein materials, flavoproteins, plastocyanin (which contains copper), three or more cytochromes (protein–haem complexes) and ferredoxin (a protein containing iron and sulphide ions) and numerous quinone–quinol systems, appear to have more probable roles as electron-transport agents than as photochemical agents, and also absorb only a small part of the total light absorbed by the photosynthetic structure.

Other pigments are found in higher concentrations, and do not have any obvious chemical role. The most convincing evidence that these are the pigments responsible for trapping the energy of light comes from comparison of the absorption spectra with the action spectrum (see Figure 2.6) for photosynthesis. From such a comparison it seems certain that chlorophyll a, carotenoids and certain other pigments are the primary absorbers. The pigments in these classes are described below.

2.51 Chlorophyll

The chlorophylls are highly coloured substances, appearing virtually black in the solid condition, and green in solution. They are insoluble in water but dissolve readily in polar organic solvents. Chemically they are neutral in solution, and possess several recognizable groups. The most obvious of these is the square ring of rings, the tetrapyrrole system of the porphyrin type. This ring system is common to the haem pigment of blood and the cytochromes. Metal ions are often found complexed in the tetrapyrrole rings: in haem it is iron, but in chlorophyll magnesium. (Other examples are copper in the red pigment of some birds' feathers (e.g. *Turaco*) and cobalt in vitamin B_{12}.) There are two carboxyl groups in chlorophyll, but both are esterified, one with methanol and one with long-chain alcohol *phytol*. Phytol is an isoprenoid; one of the distinguishing features of isoprenoids is the arrangement of side-chains and the $5N$ number of carbon atoms (here 20, $N = 4$). This is because the molecule is built up of five-carbon isopentenyl units thus:

$$C-C-C-C \qquad C-C-C \rightarrow C \qquad C-C-C-C \qquad C-C-C-C$$
$$\mid \qquad\qquad\quad \mid \qquad\qquad\quad \mid \qquad\qquad\quad \mid$$
$$C \qquad\qquad\quad C \qquad\qquad\quad C \qquad\qquad\quad C$$

Other isoprenoids include steroids, carotenoids, natural rubber and the side-chains of the many quinones of biochemical importance. Many of these are involved in photosynthesis.

For an account of the isolation procedure and chemistry of chlorophyll the reader is referred to Hill (1963); only a few details can be given here. Reference to Figure 2.5, in which the formulae of several forms of chlorophyll appear, shows that their structure is related to porphyrins, in particular, protoporphyrin IX. Magnesium occupies the centre of the square ring by displacing the acidic hydrogens. All four magnesium–nitrogen bonds are more or less equivalent, despite the conventional formula; the tetrapyrrole ring is fully conjugated so that all the bonds have some double-bond character. In the chlorophyll series, ring IV is saturated at positions 7 and 8, and in bacteriochlorophyll, ring II undergoes a similar saturation as well. The effect of these modifications is to displace the porphyrin absorption spectrum to the red. In chlorophyll *a*, there are centres of asymmetry at positions 7, 8 and 10.

Ring V is peculiar to the chlorophylls. Carbon-10 carries an acidic hydrogen atom which may be important in the photochemical reaction. There are two carboxylic acid groups, one esterified with methanol and the other with the isoprenoid alcohol phytol. Hydrolysis of the latter by the specific enzyme *chlorophyllase* (a chloroplast enzyme that only appears to act in strong aqueous organic solvents) yields *chlorophyllide*. The reaction in ethanol or methanol yields the ethyl or methyl chlorophyllide. Hydrolysis of both esters by alcoholic potash gives *chlorophyllin*. Both chlorophyllide and chlorophyllin are soluble in millinormal alkalis. Loss of magnesium from chlorophyll, which is rapid in acid solution, forms *phaeophytin* from which *phaeophorbides* are obtained by hydrolysis of the esters. In general, the chlorophylls are derivatives of dihydroporphin or tetrahydroporphin; these derivatives are termed *chlorins* when the cyclopentanone ring (ring V, the isocyclic ring) is absent, *phorbins* if it is present. *Allomerized* chlorophyll is a heterogeneous product formed by oxidation of chlorophyll in solution in air. Oxidation takes place at position 10, in ring V. Chromatography of chlorophyll extracts on silica gel, particularly in thin layers, results in the formation of multiple bands appearing similar in colour to chlorophylls *a* and *b* but moving more slowly; they are not formed to the same extent if cellulose or sucrose are used as the chromatographic bed.

Chlorophyll a has the formula I in Figure 2.5. It is universal except in the bacteria, and is usually the major pigment component of the thylakoids, by weight. From the absorption spectra in Figure 2.6, it can be seen that there are two principal peaks, in the blue and red regions. The absorption maximum in 80% acetone, the most convenient solvent for spectrophotometric estimation of the pigment, is 663 nm (see the spectrophotometric method for the assay of chlorophyll, in the Appendix); in less polar solvents it is shifted to the red, and in the virtual absence of traces of water can be found at values of around 680 nm. Aggregates of chlorophyll *a* in concentrated solutions show a small shift to the red but crystalline specimens may show a maximum at 740 nm. In the thylakoid

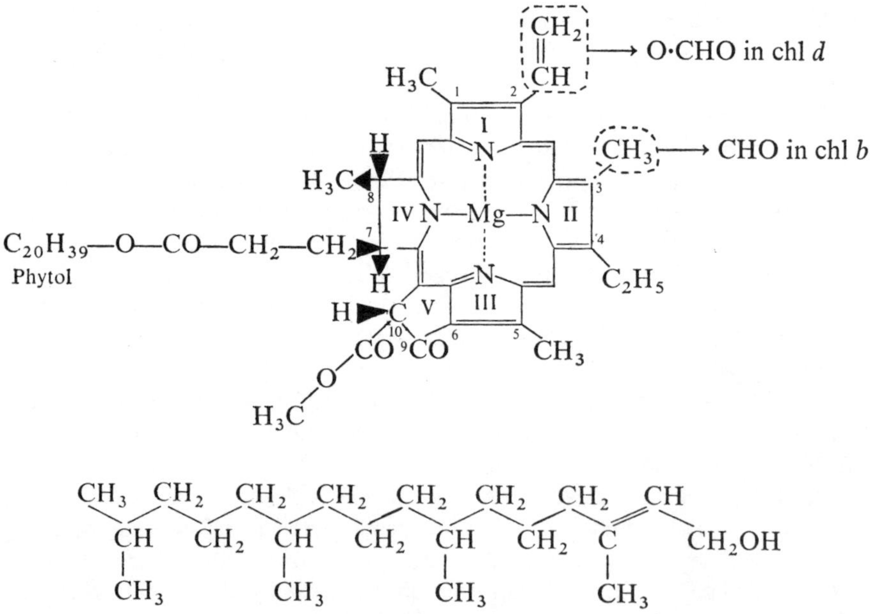

I, II, III Chlorophyll *a, b, d,*

Figure 2.5. Formulae of the principal photosynthetic pigments

Figure 2.5 (*contd.*)

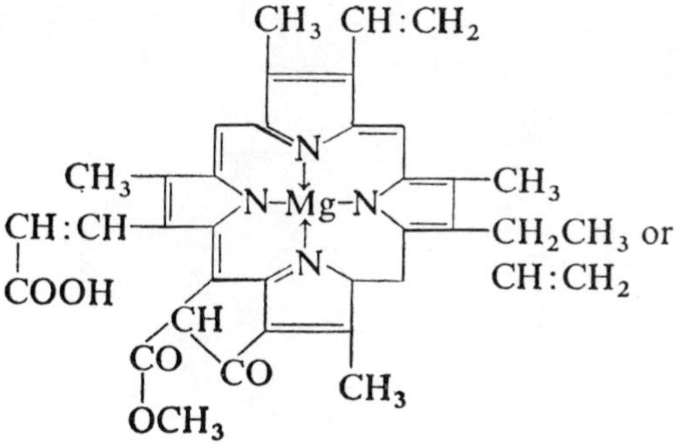

IV Chlorophyll *c*

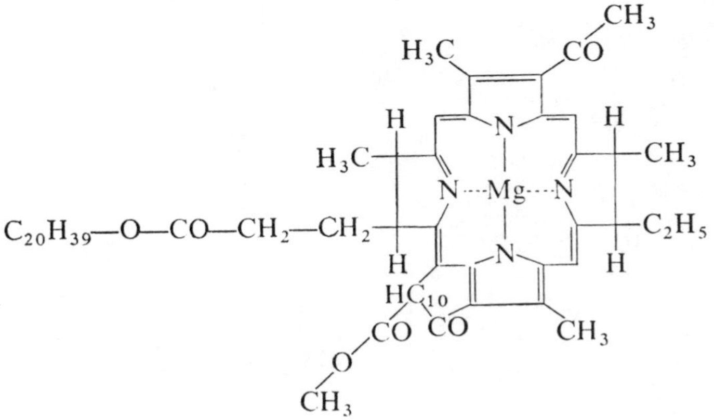

V Bacteriochlorophyll (*a*)

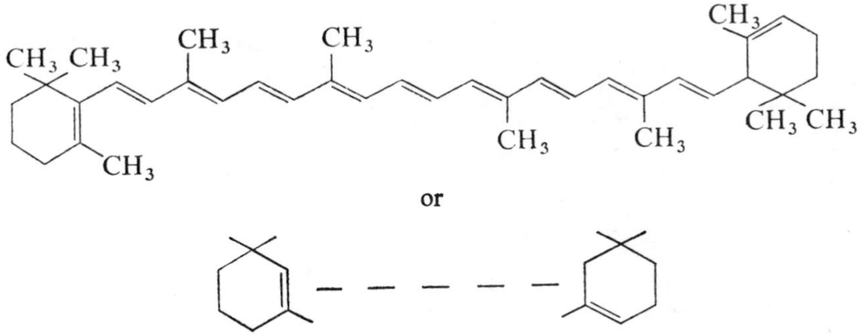

or

VI α-Carotene

Figure 2.5 (*contd.*)

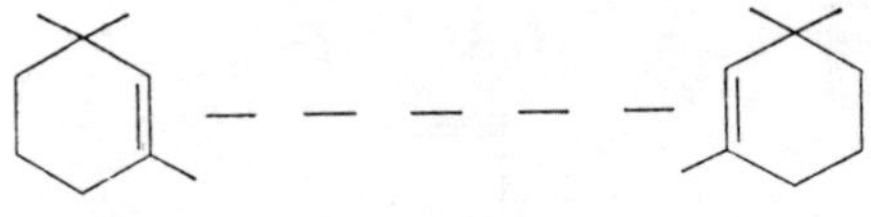

VII β-Carotene

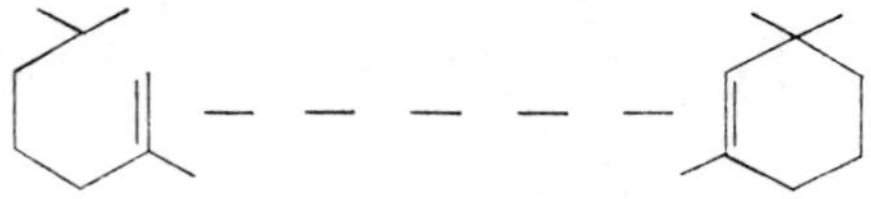

VIII γ-Carotene

IX Neoxanthin (uncertain)

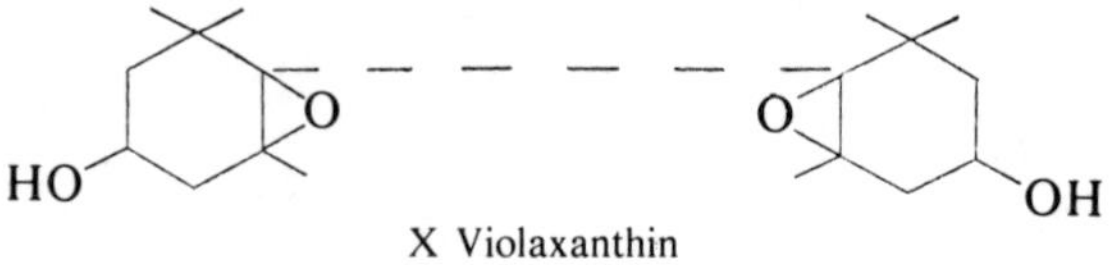

X Violaxanthin

XI Lutein

XII Antheraxanthin

XIII Zeaxanthin

Figure 2.5 (*contd.*)

XIV Spirilloxanthin

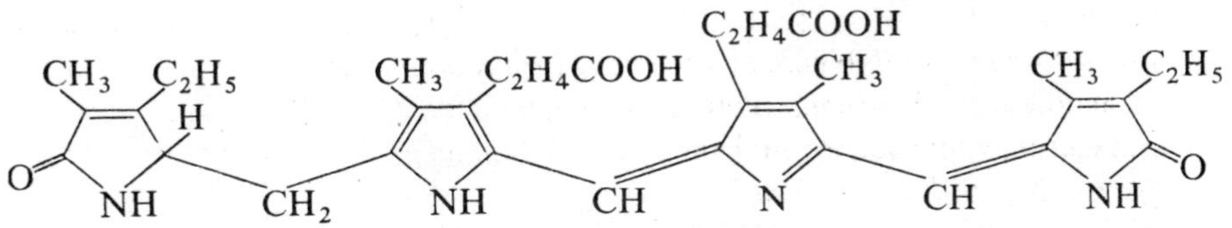

XV Probable formula for phycoerythrobilin

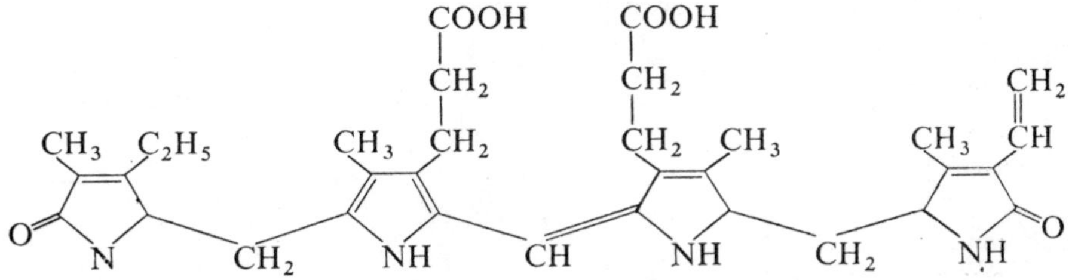

(a)

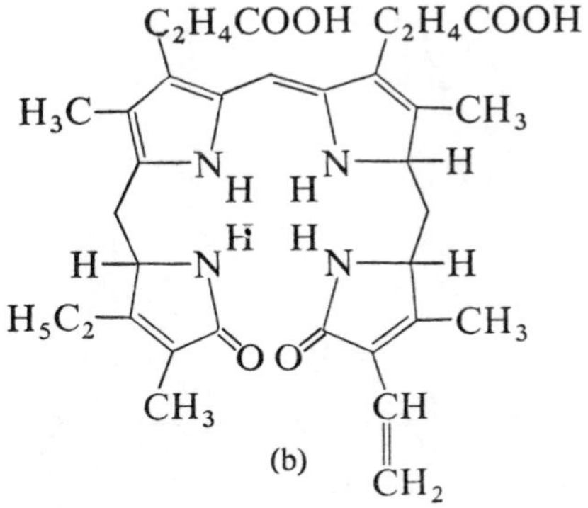

(b)

XVI Linear (a) and helical (b) formula for *d*-urobilin (see O'hEocha, 1966)

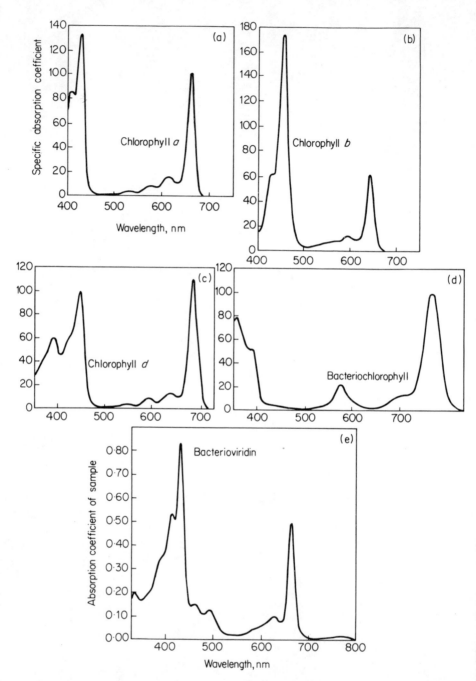

Figure 2.6. Absorption spectra of chlorophylls and other pigments. (a) to (e) from Smith and Benitez (1955), *Moderne Methoden der Pflanzenanalyse,* Vol. IV, Berlin–Göttingen–Heidelberg: Springer, pp. 142–196, with permission. (f) from Tanada (1951), *Am. J. Bot.,* **38,** 276, with permission

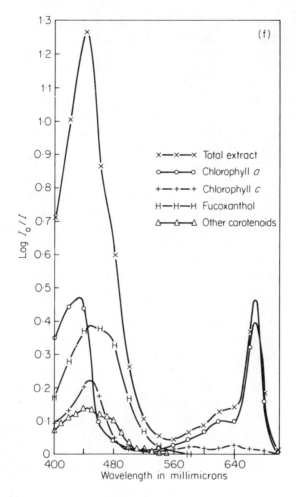

Figure 2.6 (*contd.*)

membrane, the absorption peak of chlorophyll *a* is much broader than in extracts of the same tissue in a solvent. This has been explained by supposing that chlorophyll exists in possibly five different types of environment, or aggregated states, such as might occur at different sites in a lipoprotein membrane structure. These five or more chlorophyll 'forms' are distributed between 664 and 703 nm, the principal component absorbing at 678 nm. It should be stressed that there is no reason to believe that these absorption maxima are caused by any other pigment than chlorophyll *a*, and that the differences are physical, not chemical.

Chlorophyll b has the formula II in Figure 2.5. The apparently slight difference in the formula is associated with a marked difference in the absorption spectrum (Figure 2.6) compared with chlorophyll *a*; the band in the red occurs at 645 nm in 80% acetone, and is half the intensity, while the blue-absorbing peak is slightly higher and shifted to the red. Chlorophyll *b* accompanies chlorophyll *a* in green algae and higher plants, except for a few individual species and mutant forms.

Chlorophyll c is found with chlorophyll *a* in diatoms (Bacillariophyceae), brown algae.(Phaeophyta) and dinoflagellates (Pyrrophyta). It appears to be a mixture of two compounds whose formulae are given in Figure 2.5 (IV) (Dougherty and coworkers, 1970). Ring V is present but phytol is absent, and the pigment is only soluble in polar organic solvents.

Chlorophyll d probably has the formula III (Figure 2.5). It occurs with chlorophyll *a* in some red algae (Rhodophyta).

Other green pigments, probably variants on the chlorophyll theme, are found in other algal groups.

Bacteriochlorophyll (V) differs from chlorophyll *a* in the 3,4 saturation, and also in the side chain at position 2. It is found in the purple bacteria, both Thiorhodaceae and Athiorhodaceae. In the chromatophores of these families the bacteriochlorophyll absorbs at much longer wavelengths, 800, 850 and 890 nm, compared to the pigment extracted into organic solvents which absorbs at approximately 770 nm. A second form of bacteriochlorophyll is known, bacteriochlorophyll *b*, but only in one species; it absorbs at 795 nm (960 nm in the cells).

Chlorobium *chlorophyll-660* is the principal pigment of the third group of photosynthetic bacteria, the green sulphur bacteria (Chlorobacteriaceae). This pigment was originally termed bacterioviridin. The formula is uncertain; although Fischer and Stern suggested that it was 2-acetyl chlorophyll *a*, there appears to be no methoxyl group; an ester of an alcohol other than phytol seems to be present, and the phase test, which depends on the hydrogen atom on carbon-10, is negative. If the cyclopentanone ring is absent it may mean a reappraisal of the role of this part of the molecular structure in the photochemistry of chlorophyll in general. Shifts to the red occur in the green bacteria also, the *in vivo* absorption at 740 nm altering to 665 nm on extraction of the pigment.

2.52 Carotenoids

The *carotenes* are hydrocarbons, formed by the isoprene pathway of biosynthesis, which is revealed by the carbon skeleton being a multiple of a five-carbon unit with a characteristic branching pattern. The majority are C_{40} compounds.

α-Carotene (VI) is a minor carotene component, except in the Siphonales (green algae); in all other green plants *β-carotene* (VII) is the major component. The green sulphur bacteria, on the other hand, have *γ-carotene* (VIII). Only traces of other carotenes are known to exist in photosynthetic systems. The carotenes are soluble in petroleum ether, in contrast to chlorophyll *a* which requires traces of polar solvent to be present. The xanthophylls while soluble in petrol ether are readily extracted by 90% methanol solution. A possible antioxidant role for some carotene pigments is discussed in a later chapter.

The *xanthophylls* are oxygenated carotenes, and are of a great variety. Their numbers exceed the total of chlorophylls and carotenes together, and thus only a few can be described here.

The green algae and higher plants contain principally neoxanthin (IX), violaxanthin (X) and lutein (XI). To these we may add antheraxanthin (XII) from the Euglenales, and fucoxanthin and zeaxanthin (XIII) from the brown algae. Spirilloxanthin (XIV) is the major xanthophyll in the purple non-sulphur bacterium *Rhodospirillum rubrum*.

The xanthophylls undergo interconversion during photosynthesis, at least in green plants that evolve oxygen. These pigments may protect the thylakoids from damage by excess light and oxygen. All the plant pigments are sensitive to light and oxygen when isolated, necessitating careful chromatographic procedure if confusing artifacts are not to be produced. Strain (1966) has given a useful review of these methods.

2.53 Biliproteins

Bile pigments are linear tetrapyrroles formed by the opening of the porphyrin ring. Biliproteins are proteins conjugated with a bile pigment prosthetic group. In plants, biliproteins are confined to the Rhodophyta, Cyanophyta and Cryptophyta. This is to exclude phytochrome, which is widely distributed and appears to be chiefly located in the outer membranes of cells, where it governs the responses to red and far-red illumination that underlie photoperiodism and phototropic responses. There is evidence that the algal biliproteins are aggregated on the outsides of the thylakoids (see Plate 8) giving a granular appearance. They are in general soluble, and tend to leach out during isolation of the photosynthetic organelles.

The red biliproteins, the phycoerythrins, are classified into the R, B, C and cryptomonad phycoerythrins, originally indicating their principal source. The blue pigments are the phycocyanins, with the prefixes R, C, allo and cryptomonad. O'hEocha (1966) gives a useful review of the known distribution among the algal groups. The prosthetic groups are phycoerythrobilin (XV), phycocyanobilin and phycourobilin. A helical structure (XVI) is almost certainly more appropriate than the linear scheme in all cases.

The view taken in this text, at least for the purpose of discussion, is that chlorophyll *a* in green plants and the pigments absorbing furthest into the red in the various bacteria are the primary pigments, and the carotenes, xanthophylls and biliproteins as well as the other chlorophyll varieties together are classed as *accessory pigments*. Evidence for this view will be given later.

2.6 Primary and accessory pigments

The basic theory for the operation of accessory pigments in photosynthesis is based on the observation that chlorophyll *a* (or bacteriochlorophyll) is the pigment absorbing furthest into the red; that is, that has the least energy of excitation. It is energetically feasible, therefore, for excited states of the other pigments

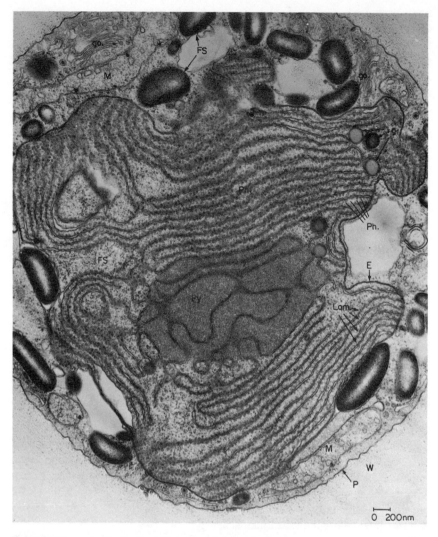

Plate 8. Electron micrograph of a section through part of a cell of the red alga *Porphyridium cruentum*. The single chloroplast which fills most of the cell is enclosed by an envelope (E) of two membranes, and contains a central pyrenoid (Py) traversed by extensions of the thylakoids. The stroma appears as a granular continuum, or matrix, supporting the thylakoid membranes (LAM) which do not form stacks or come into surface contact although they anastomose and interconnect. The surface of the thylakoid membrane bears the characteristic granules termed 'phycobilisomes' (Ph) believed to be aggregates of the biliprotein accessory pigments. Grains of floridian starch (FS), mitochondria (M), golgi bodies (go), are contained in the cytoplasm itself bounded by the plasmalemma (P) and a gelatinous sheath or wall (W). (lg)—lipid globules. Courtesy of A. D. Greenwood, Botany Department, Imperial College, London.

to pass their energy to chlorophyll *a* by the process described in section 2.4, in which molecules of chlorophyll *a* act as 'sinks' or 'traps' for some form of exciton or electron migrating from the site of the absorption of a photon. For this to be possible of course the pigments must be in suitable physical proximity. Secondly, for energy migration to occur, the absorption peak for the receiver must overlap the fluorescence emission spectral peak of the donor. The example shown in Figure 2.7(a) shows the degree of overlapping in the absorption spectra of various pigments isolated from an alga, which is sufficient for the needs of the theory. It should be noted that the fluorescence of all pigments other than chlorophyll *a* is suppressed in the living cell, indicating that they lose their excitation rapidly.

However, not all the light absorbed by all the chloroplast pigments leads to the performance of useful work. Some of the pigments may not be connected to

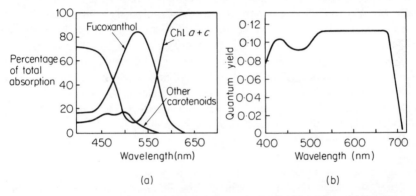

(a)   (b)

Figure 2.7. The operation of accessory pigments in *Navicula minima*. (a) To show that the light absorbed by the cells is distributed among the pigments, the percentages varying with the wavelengths. (b) To show that the quantum yield is relatively constant with wavelength, and hence that energy is being passed to chlorophyll by pigments such as fucoxanthol. From Tanada (1951), *Amer. J. Bot.*, **38**, 276, with permission

chlorophyll *a*, and all quanta face the possibility of being lost by fluorescence or heat conversion before they can reach a trap. There are two types of measurement that are of value in investigating the efficiency of pigment systems, and they are important in all areas of photobiology. The first is the *action spectrum*, which is a graph, plotted against wavelength, of the amount of measurable effect (usually, in photosynthesis, gas exchange such as oxygen production) divided by the intensity of the (monochromatic) incident light beam. The second measurement is that of the *quantum yield*, which is the amount of measurable effect divided by the intensity of light *absorbed by the sample* (i.e. the incident intensity less the transmitted and scattered intensities).

An example of the study of quantum yield is given in Figure 2.7(b). There is a middle region, 530–670 nm, in which there is very little variation of quantum yields. Therefore all quanta of these wavelengths that are absorbed have an equal chance of reaching a trap. However, reference to Figure 2.7(a) shows that in this region the proportions of light absorbed by fucoxanthol, chlorophyll *c* and chlorophyll *a* vary widely. Therefore all these pigments are efficiently connected to the trap. At shorter wavelengths this is not true—there is a greater chance of quanta being lost. One must suspect that at least some carotenoids, as well as other yellow materials in the cells, are not connected to the photosynthetic energy sink. The quantum yield also falls at longer wavelengths; this is the celebrated 'red drop' discussed in section 6.11.

It is interesting to note that in the red algae, such as *Porphyridium* (see Plate 8, p. 4G), the pigment phycoerythrin is located in granules apparently only loosely connected to the thylakoids. Nevertheless it is still able to function effectively as an accessory pigment, presumably by re-emitting light as fluorescence, at a longer wavelength which is absorbed by chlorophyll *a*. Further evidence has been obtained by studying the fluorescence of chlorophyll *a* in organisms il-luminated in regions where other pigments must absorb nearly all the light. Although chlorophyll in solution fluoresces at an intensity visible to the eye, while chlorophyll *in situ* does not, some fluorescent emission can be measured with the appropriate apparatus. Taking as an example *Porphyra*, light absorbed in the spectral range 520–560 nm, which is absorbed mainly by phycoerythyrin, promotes fluorescence in the region 680–690 nm, which must emanate from chlorophyll *a*. Moreover the intensity of fluorescence is as high or higher than that produced by direct illumination of chlorophyll *a*. Therefore in this case phycoerythrin is transferring energy to chlorophyll *a*.

The absorption spectra of the chlorophylls are sharp, and their contribution to the total absorption of the plant can be recognized easily. Other pigments have broader maxima, and identification of the components of the overall absorption or action spectra is much more difficult. The difficulty is increased by the fact that the quantities of pigments such as the xanthophylls diminish as the number of varieties increases. In the majority of species it is consequently not possible to say which of the pigments present are acting as accessory pigments. For the pur-poses of this text, it will be supposed that in general chlorophyll *b* where present has an accessory role, together with unspecified 'carotenoids' (either carotenes or xanthophylls or both).

There is no reason now to suggest that the primary pigment is other than chlorophyll *a* in green plants. However it appears that the arrangement of the chlorophyll molecules in the thylakoid, which is a very dense packing, leads to the formation of aggregates in which the absorption maximum is shifted slightly to the red, which is the reason why the absorption peak is much broader *in vivo* than when the pigments are extracted into solution. The shape of the *in vivo* spec-trum is best explained by postulating special 'forms' of chlorophyll *a* absorbing at 670, 680, 690 and 700 nm (different laboratories disagree slightly in the wavelength allocations) which can be referred to as C_{670}, C_{680} and so on. The form C_{700} can reasonably be ascribed to photoreaction I, (see section 4.45) since II is hardly active at this wavelength. Again, the fluorescence at room temperature at 695 nm, which is associated with the system of photoreaction II, probably comes from C_{690}, which must itself therefore be ascribed to that reaction. The allocation of C_{670} and C_{680} is less easy. If these forms pass their energy to one or other of the forms C_{690}, C_{700}, then in a sense they are accessory pigments. This is leading to a concept of the photosynthetic pigment system as a funnel leading from a large quantity of a short-wavelength pigment through smaller quantities of longer wavelength pigments to reaction centres containing pigment absorbing furthest to the red and present in the smallest quantities. This funnel is the photosynthetic unit, the subject of the next section.

2.7 The photosynthetic unit

The concept of the *photosynthetic unit* is basic to present-day hypotheses about the mode of utilization of light energy. Each unit is supposed to contain a *reaction centre*, where the primary chemical process of photosynthesis is carried out one molecule at a time, connected to a relatively large number of pigment molecules, any one of which may absorb a photon and pass the energy to the reaction centre. The chemical reaction is much slower than the events in the pigment mass (see Figure 2.4) and this allowed Emerson and Arnold (1932) to separate the two processes. They exposed algal cells to flashes of neon-light; the flashes were of saturating intensity, meaning that no change was observed when the intensity was increased, and they were of brief duration (less than 10^{-5} s). The rate of carbon dioxide uptake by the cells was observed during the experiment, as the flashing-frequency was varied. It was found that the photosynthetic carbon dioxide uptake reached a maximum at a flashing-frequency of some 100 s^{-1}; from this they concluded that the 'dark reaction' of photosynthesis had a half-time of $\frac{1}{50}$ s. Secondly, as the flashing-frequency was lowered from that value, although the net rate of carbon dioxide uptake fell, the *yield calculated per flash* increased to a maximum value at frequencies of the order of 1 s^{-1}. They argued that at each flash, each reaction centre was activated, and reduced one molecule of carbon dioxide. By the time that the chemical reaction was completed, the energy of the flash had died away, so that each flash resulted in one and only one molecule being reduced per reaction centre. The long dark time was necessary to ensure that all the reaction centres had completed the reaction energized by one flash before the next arrived. At this frequency, the number of molecules reduced per flash was equal to the number of reaction centres. The sample of algal cells was then analysed, and the number of chlorophyll molecules determined. Hence the size of the photosynthetic unit was found. The value given by Emerson and Arnold was about 2500 molecules of chlorophyll for the reduction of one molecule of carbon dioxide, and the values were closely similar for several species.

Since that time, the concept has been somewhat modified. The discoveries that two different kinds of light-reaction centre were required to explain some observations (see Chapter 6) and that the primary process of each was an electron transfer reaction have led to the present concept that each reaction centre moves one electron per flash, that each electron passes in turn through two reaction centres, one of each kind, and that four electrons have to make this passage during the reduction of one molecule of carbon dioxide to carbohydrate (and the production of one molecule of oxygen from water). Hence eight separate 'photoacts' are involved, so that we should expect 2500/8, say 300 chlorophyll molecules to be associated with each reaction centre. Alternatively, the two reaction centres may draw energy from a common double-unit of 600 chlorophylls.

Further evidence for the existence of such a photosynthetic unit has been obtained from the electron-microscopic observations of subunits ('quantasomes') in thylakoid membranes, by Park, which appeared to be of the size to carry 300 chlorophyll molecules. Although the subunit structure has turned out to be much more complicated, so that the term 'quantasome' has lost its precision, there is a periodicity at this order of size.

The second line of work is based on measurement of the proportion of chlorophyll to other materials of the photosynthetic apparatus. Obviously for this to be significant there must be good reason to believe that the materials chosen are an essential part of the photosynthetic process (evidence for this will be discussed later). The materials cytochrome f and plastocyanin occur at approximate proportions of one molecule of each per 300 molecules of chlorophyll a. Cytochrome f and plastocyanin are believed to act as one-electron transfer agents, implying that 300 chlorophyll molecules cooperate in the movement of one electron.

2.8 Reaction centres

The reaction centres where a chemical change occurs using the excitation energy of chlorophyll represent one of the more difficult topics of study. Not only

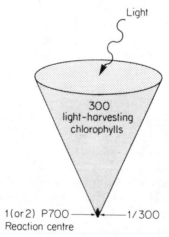

Figure 2.8. The light-harvesting and reaction centre chlorophyll of System I represented as a funnel

are the processes carried out very rapidly, but also the centres may depend on a precise arrangement of several chlorophyll molecules, which may be easily deranged when attempts are made to isolate them. It appears that there are two photoreactions in photosynthesis (possibly more) and that their mechanisms differ. The centre for reaction I (defined in section 4.54) may operate by means of a specialized chlorophyll *a* molecule known as P700 (representing about 1/400 of the total chlorophyll) which can be reversibly oxidized and reduced; one might regard the photosynthetic unit of photoreaction I as an array of some 300 'light-harvesting' or 'antenna' chlorophyll molecules connected to one or two 'reaction centre' chlorophylls (P700) (see Figure 2.8).

2.9 Summary

Light energy is absorbed by pigments of the photosynthetic system, producing an electronically excited state. Energy can be passed from an excited molecule to another pigment under certain conditions by an efficient process termed inductive resonance. Pigments such as fucoxanthin (a carotenoid), phycocyanin (a biliprotein) and chlorophylls *b* and *c* (where present) can pass their energy by this means to chlorophyll *a* and are termed accessory pigments. Some pigments do not function as accessory pigments. Approximately 2400 chlorophyll molecules cooperate in reducing one molecule of carbon dioxide, and producing one molecule of oxygen, in one flash of light. This figure is the 'photosynthetic unit'. Better, a photosynthetic unit is approximately 300 chlorophylls coupled to a reaction centre that drives one electron. If the reaction centre contains a specialized form of chlorophyll *a* (long-wavelength), the other forms of chlorophyll *a* may be regarded as accessory pigments.

Suggested further reading

Kamen, M. D. (1963). *Primary Processes in Photosynthesis*, Academic Press, New York.

Thomas, J. B. (1965). *Primary Photoprocesses in Biology*, North-Holland, Amsterdam.

Seliger, H. H. and W. D. McElroy (1965). *Light, Physical and Biological Action*, Academic Press, New York.

CHAPTER 3

Light energy into chemical energy

In the previous section a scheme was set out wherein light was absorbed by pigments, carotenes, xanthophylls and chlorophylls, producing a state of excita-. tion which was transferred to chlorophyll *a*, or bacteriochlorophyll as appropriate. The excitation was stored at some site in the chlorophyll *a* mass, and this store was in contact with a photochemical reaction centre. In this section we will consider the way in which the energy of the excitation might appear in a chemical form.

3.1 Photochemistry

On pages 26–28 four types of chemical change were listed by which an atom or molecule could lose its excitation energy. Such reactions constitute photochemistry.

The fundamental 'laws of photochemistry' upon which the foregoing principles are based, have been variously stated, but the following may provide a convenient summary.

(1) The Grotthuss–Draper law, which states that only those radiations which are absorbed by the reacting system are effective in producing chemical change. This is not to be taken as implying that all or indeed any absorbed radiation is necessarily effective.

(2) Einstein's law of the Photochemical Equivalent, which states that each molecule of a substance taking part in a chemical reaction which is a direct result of the absorption of light takes up one quantum of the radiation causing the reaction. This 'direct' result may need to be applied in such a way as to exclude reactions subsequent to the formation of the excited state. This is a similar principle to that of E. Warburg, namely that a photochemical reaction depends on the number of absorbed quanta, not on their energy content. In more modern terms, remembering that the basis of photochemistry is the excitation of an electron by a photon, we may restate Einstein's law in the form: 'The basis of a photochemical reaction at the molecular level is the activation of one electron by one photon'. This approach is well suited to the discussion of the photosynthetic unit in section 2.7.

A convenient quantity to measure in a photochemical reaction is the *quantum yield*, which is the number of molecules of a given substance formed following the absorption of one photon. (This quantity is often confused with its reciprocal, the *quantum requirement*.) The quantum yield gives an indication of the nature of the reaction. Thus very low yields imply that the chemical pathway accounts for only a small part of the total number of excited molecules, most of which are decaying (being *quenched*) by other processes. Very high quantum yields on the other hand are characteristic of chain reactions such as that between hydrogen and chlorine in which only the first step is activated by light.

3.11 Photochemistry of chlorophyll

Four kinds of photoreaction were listed in section 2.22. Of these, association and dissociation are not known for chlorophyll, neither is there any process known which might be compared to the photosensitization of H_2 dissociation by mercury atoms. All the photochemistry of chlorophyll appears to be derived from primary processes in which electrons are exchanged with other substances (oxidation–reduction reactions). Since this is in accord with what has been said so far about the primary process of photosynthesis, there has been considerable work done on the photochemistry of chlorophyll.

Chlorophyll in solution can be reversibly reduced or oxidized when illuminated in the presence of suitable electron donors or acceptors. A useful summary is given by Krasnovsky (1969), who discovered the reduction of chlorophyll to a pink compound of unknown chemical structure. The electron donor was ascorbic acid (vitamin C), and pyridine was used as the solvent, with the exclusion of oxygen. Krasnovsky and his group showed that the primary process was the formation of free-radical ions:

$$Chl^* + AH \rightarrow \cdot Chl^- + \cdot AH^+$$

that is, an electron passed from the ascorbic acid (AH) to the chlorophyll (Chl) resulting in the formation of the positive and negative charges, and both molecules then possessed an unpaired electron, which is the essential definition of a free radical. Reference to Figure 2.5 shows that there is in chlorophyll a large system of conjugated bonds, that is, alternating single and double bonds, which fuse so that the orbitals of the electrons that form these bonds cover continuously all the carbon atoms of the ring system. This conjugated system probably holds the extra unpaired electron in the ion-radical $\cdot Chl^-$. This radical stabilizes itself, forming the pink product. Chlorophyll is easily regenerated using single-ectron acceptors such as oxygen, quinones, riboflavin, methyl viologen and others. Oxygen, which possesses two unpaired electrons in the O_2 molecule, is a biradical itself; the other materials listed are able to accept single electrons, forming free radicals such as *semiquinone:*

Many reducing agents will act as electron donors to chlorophyll; in addition to ascorbic acid, cysteine, some ferrous compounds, hydroquinone and several others have been used, although in the case of hydroquinone a stable reduced product does not appear. The earlier photoproducts tend to lose their magnesium, forming phaeophytin.

Rabinowitch and Weiss in 1937 reported that solutions of chlorophyll in methanol or other solvents, illuminated in the presence of ferric chloride, became more-or-less reversibly bleached, indicating an oxidation of chlorophyll by ferric chloride. A similar photo-oxidation of the pigment has been found with oxygen, quinones, nitro compounds and methyl viologen, and as before the immediate photoproduct is an ion-radical, this time with a positive charge:

$$\text{Chl*} + \text{BH} \rightarrow \cdot\text{Chl}^+ + \cdot\text{BH}^-$$

In the presence of oxygen the unstable intermediates form peroxides, which decay with emission of light (chemiluminescence) of the same wavelength as chlorophyll fluorescence, and the pigment molecule is degraded, at ring V or elsewhere. This process, and the phaeophytin formation referred to previously, are a warning to the experimenter to protect his preparations from air and light!

It will appear later that the formation of $\cdot\text{Chl}^+$ rather than $\cdot\text{Chl}^-$ is relevant to the operation of photosynthetic reaction centres.

3.2 Primary photoproducts in photosynthesis

P700. There are two (or more) photoreactions in green plant photosynthesis, and probably in bacteria also. Reaction I (defined in section 4.54) appears to be much more like one of the photochemical systems than reaction II. As the electron acceptor is reduced, there appears in the chloroplast spectrum a diminution of absorbance at 700 nm, followed by its return. If a suspension of chloroplast fragments is carefully extracted with increasing concentrations of acetone, a stage is reached when a shoulder can be seen in the now depleted chlorophyll spectrum. This shoulder is reversibly removed by oxidizing agents such as ferricyanide, and restored by reducing agents. No pigment apart from normal chlorophyll (absorbing at 663 nm in acetone) is found when the fragments are completely extracted, and it is believed that the material absorbing at 700 nm is chlorophyll *a* modified by its environment; the bleaching is due to its oxidation. This pigment was named P700 by its discoverer, B. Kok. It is tempting to regard P700 in the oxidized state as being similar or even identical with the cation-radical formed during the photo-oxidation of chlorophyll with ferric chloride in methanolic solution. This is not proved however; P700 may turn out to be other than chlorophyll, and its oxidation may be mediated through a more immediate photo-oxidized product. A similar observation has been made with photosynthetic bacteria such as *Chromatium*, where the form of bacteriochlorophyll corresponding to P700 is known as P890. Parson, in Chance's laboratory, has shown that the oxidation of P890 precedes that of any other substance, so that oxidized P890 may fairly be regarded as a primary photoproduct. Although in the case of green plants no component has been shown to be oxidized faster than P700, there is some debate whether P700 is a necessary intermediate in the electron transport pathway.

Photoreaction II. At Witt's laboratory in Berlin a flash-induced spectral change has been reported at 687 nm which appears in less than 10^{-5} s. It is inhibited by treating the chloroplast with DCMU and reducing Q with a few flashes of light; once Q is fully reduced the 687 change no longer takes place. It is considered to be the system-II analogy of P700, and termed P680 or P690 (or chlorophyll a_{II} by Witt's group). The difficulty with P690 compared with P700 is that it has not been even partly purified, nor demonstrated using oxidizing agents. (P700 has the advantage that its reduction pathway is slow enough for the oxidized form to be easily seen. Also P700 is located on the edge of the chlorophyll absorption, making it relatively easy to distinguish.)

3.21 Application of photochemical studies to photosynthesis

The electron donors, which result in the photoreduction of chlorophyll in solution, have very little effect on the fluorescence of the pigment, whereas the electron acceptors quench fluorescence effectively. Since fluorescence comes from the singlet excitation state, it appears that photoreduction proceeds exclusively via a triplet state, and photo-oxidation at least partly via the singlet. However, nearly all photochemical reactions with dyes in solution proceed via the triplet state, for the reason that collision with a reactant molecule is unlikely during the short (10^{-9} s) lifetime of the singlet, but is possible during the much longer life of the triplet state. In fact, the kinetic analysis of such photoreactions indicates clearly that the triplet is involved in each case. The absorption spectrum of the triplet state of chlorophyll a is known, but it is only observed with any ease in solutions. In the chloroplast, although Breton and Mathis (1970) have detected the spectrum of the triplet following excitation by a saturating flash from a ruby laser, the lifetime of the triplet was only some 50 ns, since it transferred to carotenoids by an efficient triplet–triplet process. This probably accounts for the 520 nm absorption change observed by Witt and coworkers (1969a) and termed the 'valve reaction' in which very bright flashes generated more excitation in the photosynthetic units than could be used for photosynthesis, and was safely diverted to excitation of the carotenoids. The primary reactions of photosynthesis, measured by the formation of oxidized P700 and P690 (Chl a_I and a_{II}, Witt and coworkers (1969b)), take place in less than 20 ns (this is the shortest time observable with the apparatus). The fluorescence lifetime of chloroplasts is of the order of 1 ns or less. These times allow the possibility that either (i) the singlet excited state, which emits fluorescence, converts quantitatively to a triplet in 1 ns, and the triplet operates the primary processes in (less than) 20 ns, excess energy being absorbed by carotenoids in 50 ns, or alternatively (ii) that the primary processes take place in less than 1 ns, and excess energy is then lost in 1 ns, by processes which include conversion to triplets, which however are never present in significant amounts. The first scheme would be very unusual as quantitative triplet conversion is almost unheard of. The second, which is the consensus view, needs checking by means of measurements at 1 ns, which are not at present possible.

However, although the triplet state may be an unlikely intermediate in photosynthesis, the artificial photochemical systems have important applications. Thus in photoreaction I of photosynthesis, ferredoxin, an electron acceptor, is reduced, and P700, which appears to be a form of chlorophyll a, is oxidized. The photo-oxidation systems may provide a model for this process. Secondly, in one theory of the reaction centre of photoreaction II it is regarded as containing an aggregation of chlorophyll molecules, which separates charge by a photoconduction process. If the charge appears on chlorophyll molecules (the charge-traps) then these molecules are effectively anion and cation-radicals, just as in the photochemical systems. Furthermore, since these charge-traps have to initiate a chain of electron transport reactions (see Chapter 4) it is of great importance to know the redox potentials (see section 4.2) of the traps in relation to the rest of the chain, and the photochemical systems do allow an approach to be made to the measurement of the chlorophyll derivatives involved. A discussion of the estimated values of the redox potentials is given in Chapter 4.

3.3 Light energy and separation of charge

The absorption spectrum of chlorophyll in Figure 2.2 was explained in terms of the formation of singlet excited states, the bands in the red and blue forming the first and second excited states respectively. The second excited state is converted in a radiationless transition to the first. The position of the absorption maximum in the red gives an indication of the mean energy involved in the promotion of an electron from one of the vibrational levels of the ground state to one of the first excited state. The figure is obtained using the equations (see section 1.2)

$$E = h\nu = hc/\lambda$$

Taking h (Planck's constant) as $6\cdot626 \times 10^{-34}$ J s and c (the speed of light) as $2\cdot998 \times 10^{8}$ m s^{-1} we can attach energy values to the wavelength scale of Figure 2.2. The above equations give the energy in joules per photon (or per excited molecule). A more directly useful unit than the joule is the electronvolt (the work done on an electron passing through a potential rise of one volt), equal to $1\cdot602 \times 10^{-19}$ J. Since we shall be concerned in the next chapter with the redox potentials of substances taking part in photosynthetic electron transport, it is convenient to have potential measurement in volts for the energy of the excited electron of chlorophyll. On the other hand, the process of photosynthesis is directed at storing the energy of light in chemical materials, and for this purpose we find the joule per mole a more informative scale. 1 eV is equivalent to $9\cdot648 \times 10^{4}$ J mole^{-1}. Figure 3.1 sets out a suitably annotated version of Figure 1.2.

Chlorophyll *in vivo* is organized, however, into various aggregates which have absorption maxima displaced toward the red: we can assign a value for the energy of the singlet excited state in each case. The overall process of

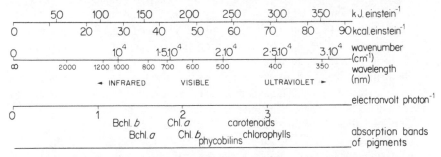

Figure 3.1. The relationship of energy and wavelength of light. Energy scales are given in terms of the energy of one einstein (the einstein is $6.023 \cdot 10^{23}$ quanta) in calories or joules, and in electronvolts per photon. Note that the spectroscopist's wavenumber scale is linear in terms of energy. The absorption ranges of the principal photosynthetic pigments are indicated below the scales. Redrawn from Clayton (1965) with permission

photosynthesis diminishes in efficiency at wavelengths on light above 680 nm and is virutally extinct at 700 nm; however this appears to be due to failure only of photoreaction II, photoreaction I being still detectable at wavelengths up to 720 nm or longer. It has been suggested that the 700 nm limit on photoreaction II

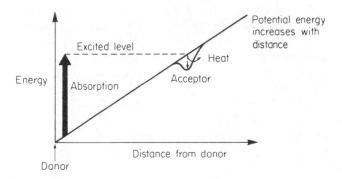

Figure 3.2. To illustrate the trapping of an electron from an excited state at some distance from its origin

is the practical limit of the 'tail' of the absorption of the chlorophyll *a* form C_{690}, and the 720 nm limit on reaction I likewise indicating the extent of the absorption of the form C_{700} (which may include P700). It should be noted that 720 nm illumination, which has an energy of 1.72 eV, can form an excited state of chlorophyll C_{700}, with energy 1.76 eV. The extra 0.04 eV is provided by the thermal energy of the pigment; the longer wavelengths only excite the 'hotter' molecules.

If the light energy that operates photoreaction I is indeed channelled through the pigment absorbing at 700 nm, then the energy available at the reaction centre is $1 \cdot 76$ eV per photon, and likewise $1 \cdot 80$ eV per photon for reaction II operating through C_{690}. This is the energy available for each photoreaction respectively. The excited electron, while still electrostatically under the influence of the positive charge of the $\cdot Chl^+$ (if that ion-radical is indeed formed) can move away some distance into the mass of surrounding material, and find a resting place in the form of an electron acceptor (see Figure 3.2).

The primary act of photosynthesis therefore establishes *a charge separation* across a certain distance. This separation of charge means that an electrical field will be set up. Suppose that the potential difference is of the order of 1 V, and the distance is of the order of the thickness of the thylakoid membrane, say 10^{-6} cm, then the field is some 10^6 V cm^{-1}, which is very large by everyday standards. One possible consequence of this is that the absorption spectra of the thylakoid membrane components will be altered by the electrical stress (the electrochromic or Stark effect). Witt's group in Berlin have indeed observed a small change in the spectra of chloroplasts (known as the 515 nm shift) rising in a time of the order of 10^{-8}–10^{-9} s. They regard this as a Stark effect of an internal field on the chlorophylls and carotenoid pigments. Moreover, their estimate of the number of charges involved in one flash corresponds to the number of reaction centres of systems I and II.

3.4 Summary

Light quanta are absorbed by chlorophyll and passed to reaction centres, where charge is separated either by loss of an electron to an acceptor (or gain from a donor), or by the more or less simultaneous appearance of opposite charges on different chlorophyll molecules separated by some distance. The former and possibly the latter concept may be clarified by work on positively and negatively charged forms of chlorophyll formed as intermediates in photoreactions in solution. P700 may be a positively charged chlorophyll ion-radical of that type. The energy taken to form the excited state of chlorophyll *a in vivo* is some $1 \cdot 8$ eV, or 176 kJ $mole^{-1}$ (42 kcal $mole^{-1}$). A spectroscopic change at 515 nm may be due to the strong electric field brought about by the charge separation.

Suggested further reading

Clayton, R. K. (1965). *Molecular Physics in Photosynthesis*, Blaisdell, New York.

Clayton, R. K. (1970). *Light and Living Matter*, Vol. 1, McGraw-Hill, New York.

CHAPTER 4

Electron transport

According to the model used for this introduction to photosynthesis, light energy is absorbed by pigments, and the excitation is channelled via chlorophyll a (or bacteriochlorophyll) to reaction centres. In green plants there are two centres, in bacteria possibly only one. At the reaction centre the energy of the excited pigment causes an electron to be transferred from a donor (which becomes oxidized) to an acceptor (which becomes reduced). Some of these primary photoreactants have been tentatively identified. In this section the principles of electron transport will be discussed, and the view will be set out in which the solid part of the photosynthetic apparatus has the role of providing an electron transport pathway in which the two light-reactions of green plant photosynthesis function in series. This view is adopted because it gives some account of most observations that have been made, and because it is held as a working hypothesis by a large majority of biochemists. There are alternative hypotheses, most of which offer more satisfactory interpretations of parts of the data, and these will be discussed in a later chapter.

4.1 Redox couples

Two compounds which are interconvertible by means of an oxidation–reduction reaction are said to be the two forms of a 'redox' substance or 'redox couple'. Although the original use of the term oxidation meant a combination with oxygen as in the formation of an oxide, most oxidations in biochemistry can be expressed as a transfer of hydrogen atoms or electrons:

$$AH_2 + B \rightarrow A + BH_2 \qquad (4.A)$$

$$X \rightarrow X^+ + e^- \qquad (4.B)$$

In reaction (4.A), AH_2 reduces B by transfer of hydrogen atoms. In reaction (4.B), half a reaction is shown in which X acts as a reductant or electron donor. A second half-reaction would of course be required since free or solvated electrons are not known to occur in biological systems. The oxidation of succinate $(AH)_2$ to fumarate (A) with the concomitant reduction of the flavoprotein enzyme succinate dehydrogenase (B $\rightarrow$ BH_2) follows reaction (4.A), and the oxidation of cytochrome c from the ferrous to the ferric form follows reaction (4.B).

4.2 Standard oxidation–reduction potentials

Redox couples can be arranged in a linear series. Their relative positions are

54

described in terms of a scalar quantity known as the *standard oxidation–reduction potential*. From these potentials one can describe and predict the changes when two couples are allowed to react. The standard potential is that which would be measured between the connections to an electromotive cell made up of one half-cell containing equimolar proportions of the oxidized and reduced forms of one couple, and a standard half-cell. Various types of standard half-cells are in use, but the ultimate standard of reference is the 'normal hydrogen electrode'. This is a platinum surface in contact both with an acid solution of unit activity (approximately 1N) and hydrogen gas at 1 atmosphere pressure. The

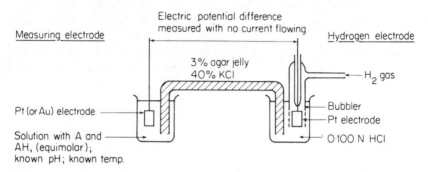

Figure 4.1. The potentiometric determination of redox potentials

reference and the 'unknown' half-cells are connected by a 'salt bridge' of saturated potassium chloride as shown in Figure 4.1. The pH of the first half-cell must be stated; pH 0 is preferred, but is not usually possible with biological material. The temperature must be defined; 25° is the preferred value. The potential is measured either by opposing it with a known potential such that no current flows, or by means of a device such as a pH meter which has an input resistance of the order of 10^{12} ohms. The sign of the potential is that of the lead from the experimental half-cell. This is the 'European' convention. The symbol is E_0 or E_0'. If the pH is other than zero, the symbol E_0' is used, with an indication of the actual pH value. At pH 7 the standard potential E_0' of the hydrogen electrode is diminished by 7×60 mV to -0.420 V at 25° (see Equation (4.3)).

If the ratios of the oxidized and reduced forms of the couple are varied, the potential E is no longer the standard value (E_0), and is given by

$$E = E_0 + \frac{RT}{nF} \ln \frac{[\text{oxidized form}]}{[\text{reduced form}]} \tag{4.1}$$

In this equation n is the number of equivalents per mole, for example two in reaction (4.A) and one in (4.B). F is the Faraday, equal to 96 480 coulombs per gram-equivalent, R is the gas constant, 8.314 joules deg^{-1} mole^{-1}, and T is the absolute temperature. Then at 25°C and using decadic logarithms

$$E = E_0 + \frac{0.05915}{n} \cdot \log \frac{[\text{oxidized form}]}{[\text{reduced form}]} \tag{4.2}$$

The numerical factor increases by 0·0002 per degree over the physiological range. For most purposes 0·05915 V can be written 60 mV or 0·06 V.

If a reductant of the AH_2 type can dissociate as an acid:

$$AH_2 \rightarrow AH^- + H^+$$

$$AH^- \rightarrow A^{2-} + H^+ \tag{4.C}$$

then the redox potential will vary with the pH. This dependence is numerically similar to Equation (4.2)

$$E_0' = E_0 + a \cdot \frac{0·05915}{n} \cdot \log |H^+|$$

$$= E_0 - a \cdot \frac{0·05915}{n} \text{ (pH at 25°C)} \tag{4.3}$$

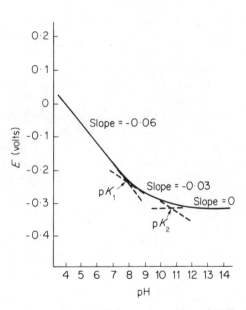

Figure 4.2. The dependence of E_0' on pH for anthraquinone-2,7-disulphonate. From E. H. Mahler and H. R. Cordes (1966), *Biological Chemistry*, Harper and Row, New York, Fig. 5.6, p. 208, with permission

The term a/n in Equation (4.3) is the number of hydrogen ions released per electron transferred, so that with a dibasic acid of the type AH_2, a/n has three values: zero when the pH is considerably above the second pK, $\frac{1}{2}$ for pH values between the two pK values (provided these are separated by at least 2–3 pH units), and unity for pH values below the first pK. Within about one pH unit on either side of each pK value, a/n has a varying fractional value so that a smooth curve results for E_0' plotted against pH. A typical curve is sketched in Figure 4.2.

The above treatment of standard potentials needs a further qualification with materials such as cysteine, which in the oxidized form (cystine) is dimerized. This means that not only does the term 'equimolar proportions of oxidized and reduced forms' cease to mean the same as '50% reduction', necessitating an *ad hoc* specification of the 'standard' cysteine-cystine couple, but also that a squared term appears in Equation (4.1)

$$E = E_0 + \frac{0 \cdot 06}{n} \log \frac{|\text{cystine}|}{|\text{cysteine}|^2}$$

Enzymes which depend on the $(SH)_2/S-S$ balance for their stability are therefore sensitive to changes in concentration (see p. 182).

From Equation (4.1), a tenfold change in the proportions of oxidized to reduced species causes a change of $0 \cdot 06/n$ in the potential. This sets practical limits on the oxidizing power of a redox material in the same way as there are limits to the buffering range of a weak acid in the presence of its salt. Just as it is convenient to formulate a pH buffer within the range of 1 pH unit on either side of the pK of the weak acid, so that the ratio of acid to salt is within the range $0 \cdot 1$–10, so the corresponding range for a redox buffer is $(0 \cdot 06/n)$ V. With present day methods of analysis, a material can be said to be 'completely' oxidized or reduced when the ratio exceeds 10^2 or 10^3 to one. If therefore a gap of $0 \cdot 24$ V separates the standard potential of two redox materials, the lower can have no significant oxidizing action upon the upper if n is unity for both; if n has a higher value, the necessary gap is correspondingly less. For practical purposes, a couple which has a higher redox potential will oxidize a couple with a lower one. Where the difference is less than $(0 \cdot 24/n)$ V, a measurable equilibrium will be set up —provided any necessary catalyst is present.

Provided that conditions of pH, temperature and concentration are considered as and where appropriate, the standard oxidation–reduction potential of a redox material, and the relative standard potentials of a group of redox materials, are of the greatest value in elucidating the biochemical role which such groups play.

4.3 Coupled oxidations

If two separate redox couples are mixed and they react, they must reach a common potential by adjustment of the ratios of the oxidized to reduced species of each couple, each according to Equation (4.1). If the two couples do not react, they may yet be brought to a common equilibrium potential by the addition of a third redox material with which they both react. In this case the oxidation or reduction of one material by a second is said to be coupled by the third.

For substances in free solution, it is rare to find systems of more than four such redox systems coupled together. In such a system, the standard potentials must either increase from left to right, or else, if there is a step in the reverse direction, the considerations of the previous section must limit it to the order of $0 \cdot 1$–$0 \cdot 2$ V.

One example is the reaction

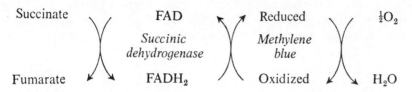

In the above example the standard potentials are (at pH 7): succinate–fumarate, 0·03 V, FAD–FADH$_2$ (in succinic dehydrogenase), approximately 0 V; methylene blue (oxidized–reduced), 0·01 V, and oxygen–water, +0·82 V.

For the sake of brevity, redox couples will be referred to from now on more briefly; the potential of cytochrome f will stand for the potential of the

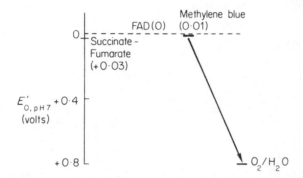

Figure 4.3. Diagrammatic representation of an electron transport process, using a scale of standard redox potentials

oxidized–reduced cytochrome f couple. There is seldom more than one redox reaction associated with each substance that we shall be considering.

In the above example electron transfer took place from a couple of lower standard potential (succinate–fumarate) to one of higher potential (oxygen–water). This is the natural or 'thermodynamic' direction (see Figure 4.3). Electron transfer in the reverse direction requires a supply of energy, either to be supplied by adjustment of the concentration ratios of the reactants so as to bring their actual potentials into the proper order, or by introducing a simultaneous additional chemical change. Chemical changes that are 'coupled' to an oxidation–reduction reaction are common in enzyme-catalysed biochemical processes, and often involve the formation or breakdown of nucleoside triphosphates:

$$AH_2 + B + ADP + P_i \rightarrow A + BH_2 + ATP \qquad (4.D)*$$

The coupled reaction has a characteristic *standard free energy change* ΔG_0, the

---

* ADP represents adenosine diphosphate, P$_i$ represents the orthophosphate ion, ATP represents adenosine triphosphate. The water formed in the condensation is not usually shown in the equation; there is also an uptake of hydrogen ions, which is not shown either.

actual free energy change, ΔG, depending on the concentrations of the reactants and products, and on the pH. This free energy change superimposes on the potential difference, E, for the oxidation–reduction reaction, a potential ΔE related to the free energy change by the expression

$$\Delta G = -n\mathrm{F}\,\Delta E \quad \text{where F is the faraday,}$$

96 480 joules mole^{-1} (23 088 cal mole^{-1}). In example (4.D), taking a free energy change of -33 kJ mole^{-1} (-8000 cal mole^{-1}) for the formation of ATP under physiological conditions, the change in the redox potential difference for the couples AH_2–A and B–BH_2 is $0\cdot173$ V (in this example $n = 2$).

4.4 Electron transport in biological systems

In living cells, there are solid systems in which oxidation–reduction reactions are believed to occur by means of several steps in series, each involving a specific

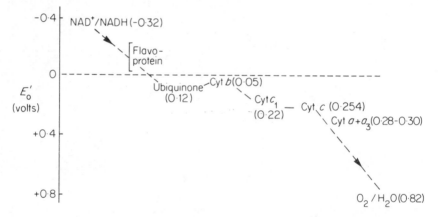

Figure 4.4. Representation in the same way as in Fig. 4.3 of the electron transport in the mitochondrial respiratory chain. By permission of M. Klingenberg

redox couple. Examples are the inner membranes of mitochondria, the membranes of the endoplasmic reticulum obtained as 'microsomes', the outer and inner membranes of the nuclear envelope in certain cells, chloroplast thylakoids in green plants, thylakoids in the blue-green algae, and membrane preparations from bacteria which may have both oxidative and photosynthetic roles. Probably the system which has been studied at the greatest intensity is the mitochondrial pathway of oxidative phosphorylation, in which reduced nicotinamide adenine dinucleotide (NADH) and succinate are oxidized, oxygen is reduced to water, and ADP is phosphorylated to ATP in fresh preparations. For details of the mitochondrial 'respiratory chain' reference may be made to Klingenberg (1968). Figure 4.4 reproduces a diagram summarizing the process, in the same form as Figure 4.3. The reactions of the chloroplast will later be drawn on such a diagram for comparison (Figures 4.7 to 4.10).

The inner membrane of the nuclei of certain cells appears to possess a respiratory chain of the mitochondrial type; the outer membrane and the microsomal system possess pathways of electron transport which are less well understood. For a discussion of these the reader is referred to Strittmatter (1968). It is not established whether the cell membrane (cell envelope) has electron transport pathways, but the same membrane in bacteria, which appears to be the only membrane system present, is believed to carry out the functions of the mitochondria of eukaryotic cells. Smith (1968) gives a useful review of this topic.

It happens that the redox components of these membrane systems —transition-metal compounds, quinones, flavoproteins and so on—have different absorption spectra according to whether they are in their oxidized or reduced states. Provided that the biological material is sufficiently transparent to light (visible or ultraviolet) in the region of the difference, the changes in the redox state of a given component can be observed directly by a spectrophotometer during the operation of the electron transport pathway. By observing two or more components simultaneously in this way, the order in which they react can be established. Reference has already been made to this in section 3.2.

Often the first indication of the presence of a redox component in a biological structure is the observation of a change in the absorption spectrum during an oxidation–reduction reaction. The membrane material is then fractionated and the unknown material purified and characterized. Redox materials seem to fall into relatively few families, and the chemical nature of many redox intermediates can be inferred directly from the difference spectrum of the oxidized and reduced states of the original material.

After the discovery of the electron transport system in the mitochondrion, other sites were discovered as described above. The concept of electron transport includes three features: first, there must be several, usually at least three, redox components present; second, most if not all of the intermediary components must be bound together in one 'solid' matrix such as a membrane; and third, it must be shown that electron transfer takes place in a sequence from one to the next.

These features are found in photosynthetic systems.

4.5 Electron transport in photosynthesis

Fresh preparations of chloroplasts from peas and other plants are capable of taking up carbon dioxide when illuminated (whether as the dissolved gas or as bicarbonate ions will be discussed later) and forming carbohydrate; oxygen is evolved in this process. This provides a reason, in this text, for concentrating biochemical attention on the chloroplast as the site of photosynthesis. For a treatment of the problems involved in the maintenance by the organism of the necessary conditions for the chloroplast to do this *in vivo* reference may be made to Rabinowitch (1945, 1951). We will be concerned with the control that the cell undoubtedly exerts over the photosynthetic process in the chloroplast. However for the present it is sufficient to state that the chloroplast carries out the process described by the equation, first derived by de Saussure:

$$CO_2 + H_2O \rightarrow (CH_2O)_n + O_2 \qquad (4.E)$$

This process is clearly the reverse of the overall equation describing the respiration of carbohydrate:

$$(CH_2O)_n + O_2 \rightarrow CO_2 + H_2O$$

Equation (4.E) is an oxidation–reduction reaction, and all the component reactions which it summarizes are located within the envelope of the chloroplast in plant cells.

It is widely accepted that the metabolism of carbon, from carbon dioxide to carbohydrate, takes place in the solution (stroma) phase of the chloroplast. The evidence is not complete, and alternative points of view will be discussed in a later chapter. For the present purpose, however, the above hypothesis will serve. The carbon metabolism requires a supply of reduced NADP and ATP besides carbon dioxide, enzymes, cofactors and catalytic substrates. There is some analogy here with the oxidative process of the mitochondrion, which takes place mostly in the stroma (matrix), and the NADH is oxidized by the cristae, which produce ATP. It is established that given only one soluble intermediate, the protein ferredoxin, the thylakoids of the chloroplast reduce NADP and produce oxygen and ATP. In the analogy, the cristae of the mitochondrion correspond with the chloroplast thylakoids.

For the rest of this chapter we shall examine the redox materials of the chloroplast, their properties and location. A scheme can be drawn showing possible reaction pathways, but the evidence for it will be discussed later.

4.51 The redox components of photosynthetic systems

Small molecules. Nicotinamide coenzymes. NADP* (Figure 4.5.I) is the chloroplast coenzyme. NAD is not known to occur in chloroplasts of green plants. In the blue green algae, where there is no distinction between 'stroma' and cytoplasm, the roles of NAD and NADP in photosynthesis are uncertain. In bacteria it is probable that NAD is important.

Quinones. The principal quinones are the plastoquinones (Figure 4.5.II) and tocopherylquinones (Figure 4.5.III). They are soluble only in hydrocarbons or absolute ethanol, where their redox potentials, measured indirectly, may have only an approximate relevance to their environment *in vivo*. *Lipoic acid* is present, but is seldom assigned a role. *Manganese* ions, or an easily dissociated complex of manganese, can be detected and has an ascribed role.

Flavins. Flavin mononucleotide (figure 4.5.IV) is used as a reagent but does not appear to occur free; neither does flavin dinucleotide (FAD).

---

* The convention to be followed here is that the acronym NAD or NADP does not indicate the redox state: the forms NAD^+, NADH, $NADP^+$, NADPH indicate the oxidized and reduced states respectively.

Protein conjugates. Conjugated proteins play a large part in electron transport, and biological oxidations in general. The oxidoreductase enzymes activate specific metabolic substrates and transfer electrons or hydrogen atoms between them and other metabolites or coenzymes. Some coenzymes are sufficiently tightly bound to their protein that they can be regarded as prosthetic groups. Not all these proteins are enzymes; if such a conjugated protein can accept (or donate) electrons from a wide variety of sources, then activation of specific substrates is no longer part of the mechanism, which is what many biochemists would consider essential to the definition of an enzyme. It is with this group that this section is concerned.

The redox properties of these conjugated proteins lie in the nature of the prosthetic groups, and the proteins can be grouped in this way. The prosthetic group, when separated from the protein, has a characteristic redox potential, but *in situ* this is raised or lowered by the influence of the protein. Thus the flavin nucleotides FMN and FAD have a standard redox potential E_0' (pH 7) at 25° of -0.185 V, whereas the *flavoproteins* can have values between zero (succinate dehydrogenase) and -0.38 V (ferredoxin-NADP oxidoreductase). While many flavoproteins can be isolated from leaves of plants, it is doubtful whether so many are concerned with photosynthesis or are even chloroplast components. The yellow enzyme ferredoxin-NADP oxidoreductase (NADP reductase) mentioned above is the only flavoprotein with a role in the scheme of photosynthesis being presented here. It is located in the thylakoid membrane, but is slowly leached out *in vitro*. The molecular weight is 40 000; there is one mole mole^{-1} of FAD and E_0' (pH 7) lies between NADP (-0.32 V) and ferredoxin (0.42 V).

Copper atoms or ions are often found in conjugation with proteins, such as haemocyanin in invertebrates, or the polyphenol oxidases. Usually there is an intense blue colour, indicating that d–d transitions which were 'forbidden' in the simple ions, have become 'allowed' in the complexed state. The standard poten-

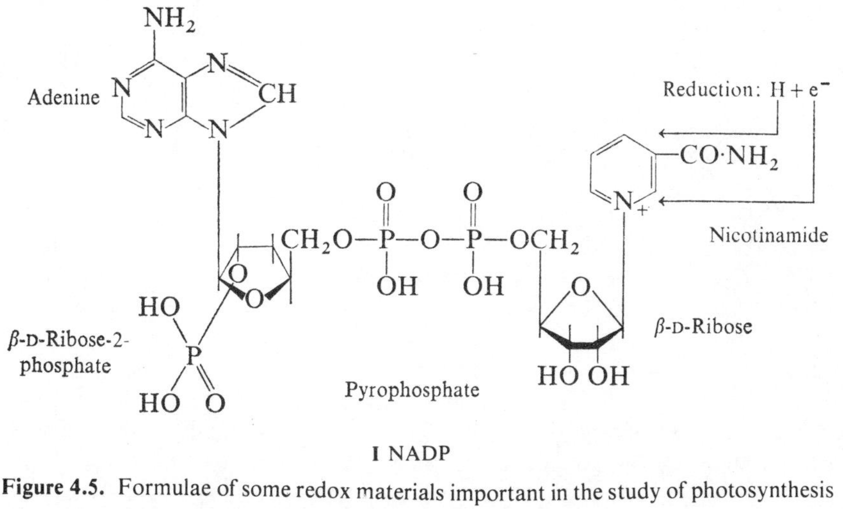

I NADP

Figure 4.5. Formulae of some redox materials important in the study of photosynthesis

Figure 4.5 (*contd.*)

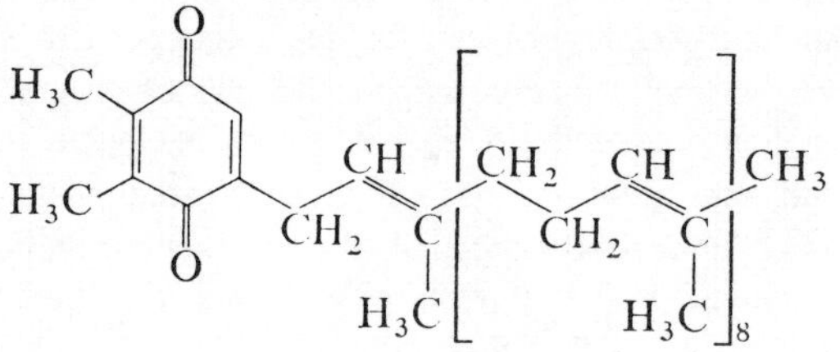

II Plastoquinone (Kofler's quinone; Q 254; PQA-45)

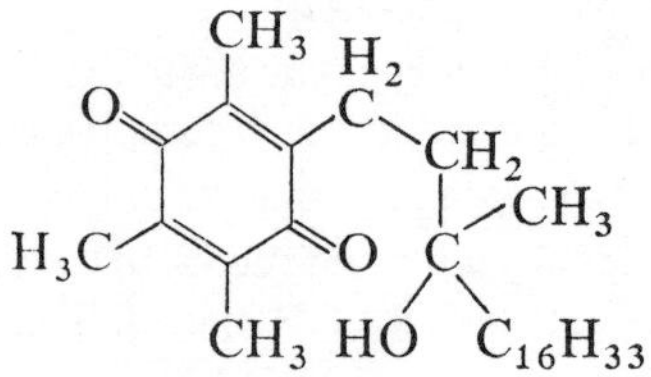

III α-Tocopherylquinone

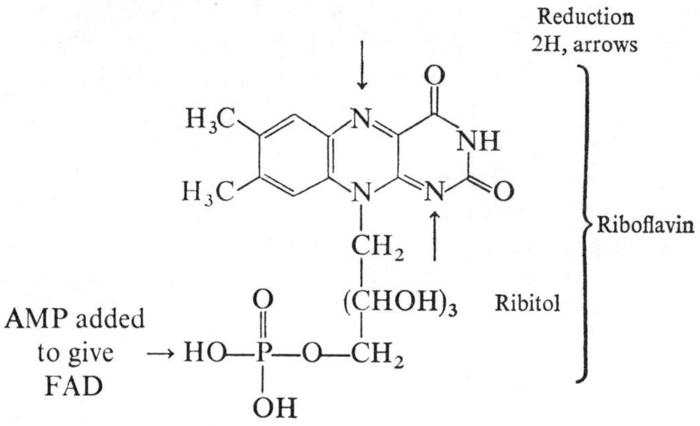

Reduction
2H, arrows

Riboflavin

(CHOH)$_3$ Ribitol

AMP added
to give   → HO—P—O—CH$_2$
FAD

IV FMN

tials are closer to that of the cupric–cuprous couple, the scatter being some
0.35 ± 0.05 V. In the chloroplast the protein *plastocyanin* (molecular weight
10 500, one atom of copper) has an intense blue colour, a potential of $+0.37$ V,
and apparently an important electron transport role in the thylakoid membrane.

'*Non-haem' iron* has recently been found to be a constituent of several enzyme
systems, and is characterized by absorption in the blue end of the spectrum,
release of ferrous iron by acid, and usually release of hydrogen sulphide at the
same time. The nature of the prosthetic group is not certain, although a bridged
complex of iron ions and sulphur atoms seems likely. The observed range of 'NHI'-
proteins covers a potential span of from zero (in the succinate dehydrogenase

complex) to the very low potentials of the *ferredoxins* of bacteria and plants. Plant (spinach, *Chlorella* and parsley) ferredoxins are located within the chloroplast, and have potentials of the order of -0.420 V at pH 7 and 25°. They are red to red-brown in colour and have molecular weights of 12–13 000, with two atoms each of sulphur and iron. Either ferredoxin is only very loosely bound to the thylakoid, or else it is a stroma protein; unless care is taken the preparation of chloroplasts always involves some damage to the envelope, and ferredoxin is lost.

P700 is presumed to be a protein conjugate, on somewhat inadequate evidence. The measured potential is $+0.430$ V. There seems little point in comparing it with the potential of $+0.645$ V observed for chlorophyll oxidation by ferric chloride (in ethanol) until more is known of the nature of the reaction.

Haem conjugates. Of the three groups into which haem-proteins are usually divided, the haemoglobin group is not represented in the chloroplast. Of the catalase–peroxidase group, catalase is certainly represented, in both stroma and thylakoid, although the quantity is much less than in the cell peroxisomes. Nevertheless there is sufficient to cause a fast evolution of oxygen in the presence of 1 mM hydrogen peroxide. Catalase is only known in the ferri-haem state, although it can be reduced after partial denaturation. Although it is the only enzyme known which can evolve oxygen, and although it may possibly form redox couples of very low or very high potential, no direct role has been found for this enzyme in the electron transport system. Both catalase and peroxidase are found in bacterial chromatophores.

The remaining class of haem-proteins comprises the *cytochromes*. Two types are represented in thylakoids, cytochromes *b* and *c*. The *b* cytochromes have been only recently isolated from thylakoids, and their identification *in situ* has been by means of spectroscopic observation of the material, usually after removal of the chlorophyll. Cytochrome b_6 has a potential of 0.00 V, and a characteristic alpha-band in its spectrum at 563 nm. There is some reason to believe that it may have two components, one of which may absorb at 559 nm. Cytochrome b_3, with a potential of between 0.058 and 0.260 V, was originally isolated from plant microsomes, but it has been claimed that it exists in chloroplasts as well; it absorbs at 559 nm. A third cytochrome absorbing at 559 nm has been shown by Bendall (1968) to have a potential of $+0.37$ V, which is remarkably high. It was extracted by Garewal and Wasserman (1974) and shown to be a *b*-type cytochrome, ('*b*-559'). (The chemical test for *b* cytochromes is based on their dissociation in acid-acetone, releasing protohaem IX (see Figures 4.6.I and 4.6.II).) b_6 has been characterized in the same way. Cytochrome *b*-559 undergoes a change in potential under treatment that mildly disrupts the chloroplast, reaching a value of approximately zero volts. The two forms are distinguished by the symbols *b*-559 HP and *b*-559 LP.

The *c* cytochromes are represented in chloroplasts in higher plants by cytochrome *f* (*f* standing for *frons*, a leaf), which has been isolated; it has a molecular weight of 120 000, two haem groups, and a potential of $+0.370$ V.

The alpha-band is at 555 nm. In the green algae a very similar cytochrome is known as C552. Cytochrome *f* has been allocated the designation c_6.

Preparations of bacterial chromatophores contain cytochrome c_2, which has a potential of $+0.33$ V, or other *c*-type cytochromes. The range of molecular weights and potentials is great. The occurrence of the *b*-type cytochromes was

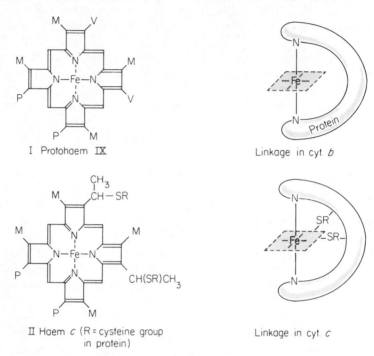

I Protohaem IX

Linkage in cyt. *b*

II Haem *c* (R = cysteine group in protein)

Linkage in cyt. *c*

Figure 4.6. The binding of haem in cytochromes of types *b* and *c*. V = vinyl; M = methyl; P = propionic acid.

thought to be limited to the Athiorhodaceae (see Table 8.1). A 'hybrid', '*Rhodospirillum* haem protein, RHP' has been isolated from both *R. rubrum* and *Chromatium*. It has intermediate cytochrome properties, and a potential of approximately zero. The purple bacteria also contain cytochromes, C422 and C423·5, which have been mentioned previously; they are believed to lie close to the reaction centre.

4.52 Cyclic electron transport

In section 3.2 the qualified assertion was made that the result of the absorption of light by the chloroplast resulted in the formation of oxidizing and reducing agents as the first detectable step. If these entities were allowed to react with each other, either directly or by the mediation of a number of redox materials, a steady state could be visualized in which electrons were driven round a chemical circuit by the energy of light. One might detect a change in the redox balance of a given component when the light was switched on or off, but in the steady state there would be no release of oxidizing or reducing equivalents to the surrounding space (see Figure 4.7).

The value of such a cyclic process lies in the observation that passage of electrons through biological chains of redox substances is often associated with the formation of the energy-storage compound ATP. A typical case is that of the mitochondrion where the electrons originate from the oxidation of respiratory substrates such as succinate, and leave to form water from atmospheric oxygen. Hypotheses concerning the mechanism of the coupling of electron transport to

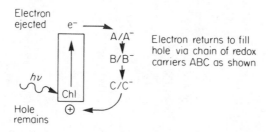

Figure 4.7. To illustrate the principle of cyclic electron flow, using the style of Fig. 4.3. (Compare with Figs. 4.8 and 4.9)

phosphorylation of ADP are many, and their testing is a problem of general biochemistry. For the present, let it be assumed that whatever process holds in one case holds for all.

4.53 Non-cyclic electron transport

The life of the green plant depends on the reduction of carbon dioxide to carbohydrate by means of light energy. It is clear that in this case a reducing agent is being released by the photosynthetic process. At the same time oxygen is evolved, so that it is clear that non-cyclic electron transport takes place.

Unless the separated oxidizing charge resulting from the illumination of chlorophyll passes directly to water, either one or a chain of redox substances must be involved to couple the electron transport. This hypothetical chain is usually regarded as passive, and hence the standard potentials of the components must be around or above the potential of the hydroxyl-ion–oxygen reaction (+815 mV at pH 7). In Figure 4.8 the light-induced oxidant Y_{11} is shown at an arbitrary potential, more positive than 815 mV.

The Hill Reaction

There must be a light-induced reductant corresponding to the oxidant above. To complete a path for non-cyclic electron flow an electron acceptor must be available. Before the discovery of the natural acceptor, ferredoxin, other oxidizing agents were known, such as complexes of ferric iron (the uncomplexed ion Fe^{3+} is not soluble at physiological pH values), benzoquinone and indophenol dyes. In the presence of one of these oxidants, chloroplasts or chloroplast fragments will evolve oxygen when illuminated. This is the *Hill Reaction*, and the oxidants are *Hill oxidants*. It is subject to the 'red drop', that is, a decline in quantum yield at wavelengths above 680 nm virtually reaching zero by 700 nm.

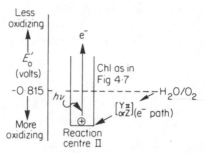

Figure 4.8. Reaction-centre II and oxygen production

There is apparently another passive electron transport chain connecting the primary, light-induced reductant (X_{II}) with the site at which the Hill oxidants act. Plastoquinones are likely to be involved. Roles for the cytochromes b_6 and b-559 at this point have been suggested, but must be considered unlikely. Spectroscopic entities such as X-320 and C-550 may be related to X_{II} (see Chapter 8).

The Hill reaction proceeds at a greater rate if the chloroplasts are supplied with ADP, phosphate and magnesium ions. ATP is then formed. This indicates that phosphorylation is coupled to the electron transport of the Hill reaction (as opposed to a possible recombination of oxygen and reductant by some 'mitochondrial' system). Ammonium salts, and amines, release the coupling analogously to uncoupling agents like dinitrophenol in the mitochondrion. (These uncoupling agents often inhibit the Hill reaction in the chloroplast at higher concentrations.) Phosphorylation of ADP accompanied by the oxidation and reduction of substrates is termed '*non-cyclic* photophosphorylation'.

There is evidence that manganese and chloride ions are required for the Hill reaction; this will be discussed later.

The reduction of NADP. The bisphosphoglyceric acid reductase (glyceraldehyde-3-phosphate dehydrogenase, EC.1.2.1.13) in the chloroplast stroma is specific for NADP. It is believed that this is the principal site of reduction for the assimilated carbon dioxide in photosynthesis, although some objections must be considered later. The thylakoid electron transport system can be said to terminate with the reduction of NADP. This again shows a formal analogy with the mitochondrion, since in each organelle electron transport is between oxygen and a nicotinamide coenzyme. The enzyme responsible for this reduction is NADP reductase, the flavoprotein described on page 61. This flavoprotein will accept electrons from ferredoxin, and this in turn is reduced by thylakoid preparations in the light. The reduction of ferredoxin alone is not easy to demonstrate if oxygen is evolved or if air is present as the ferredoxin is reoxidized; an artificial electron-donor, ascorbate plus dichlorophenol indophenol, can be used anaerobically. There seems little doubt that the final sequence of the electron transport pathway is represented by: reaction centre I: X_I: ferredoxin, NADP reductase and NADP (Figure 4.9). The symbol X_I both identifies the position in

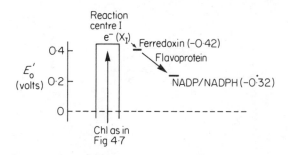

Figure 4.9. The NADP⁺-reducing system; compare with Figs. 4.7 and 4.8

simple diagrams of the zigzag scheme, and provides a label for the hypothetical intermediate acting at that site.

4.54 Two light reactions

The reduction of Hill oxidants does not always correspond with the reduction of NADP by the above sequence, and for various reasons, set out and discussed later, the idea of two light reactions in series has been introduced. That labelled System I reduces NADP as described above, and generates an oxidant, which can be labelled Y_I. System II oxidizes water and generates a reductant, which we have already labelled X_I, and which is oxidized by Hill reagents. It is necessary to suppose that X_I can be oxidized by Y_I either directly or via a chain of redox intermediaries. This scheme is summarized in Figure 4.10. It is tempting to try to place all the pathways on a common diagram, so as to produce a 'metabolic map' for electron transport. At present the combined uncertainties of each path tend to arouse confusion. Thus there is no guarantee that the site of ATP formation is the same for cyclic and non-cyclic pathways. Figure 4.10 also shows in parentheses positions of redox materials which are at present subject to criticism. Evidence which affects the placing and position of each redox material on a figure such as Figure 4.10 (if such a figure represents the state of affairs at all) will be reviewed later.

4.55 The bacterial system

Illumination of the chromatophores of *Rhodospirillum rubrum*, a purple non-sulphur bacterium, causes the oxidation of a cytochrome, C422. This reaction takes place at temperatures of 77°K indicating that it is an electron movement,

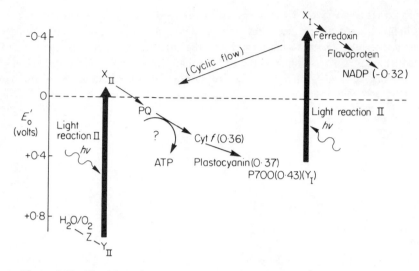

Figure 4.10. The 'zigzag' or 'Z' scheme for electron transport in green plants

not involving movements in atomic nuclei, and hence that it is a primary photoreaction. Bacterial cells do not generate oxygen, and the photosynthetic process is anaerobic. ATP is produced, and this must be directly coupled to the photosynthetic electron transport. When provided with the usual coupling agents, chromatophores carry out cyclic photophosphorylation.

Clearly, since photosynthetic bacteria assimilate carbon using light energy, non-cyclic electron transport must be taking place. The point has been made however that given a supply of ATP from the cyclic process, the non-cyclic electron transport might be a 'dark' process similar to the reduction of NAD by succinate in mitochondria. In mitochondria the process is known as *reverse electron transport*.

Assuming that the non-cyclic electron transport system is analogous to that in the chloroplast, then measurements of the rate of reaction with light intensity indicate that at least two quanta are required per electron. The bacterial system operates between much closer values of redox potential, and it is not easy to see whether two quanta are necessary. ATP is formed during the non-cyclic process. The status of this apparent non-cyclic photophosphorylation will be discussed in Part II, Chapter 8.

The relations between the electron transport pathway and the carbon metabolism of these bacteria are capable of great variation. Hydrogen donors encountered are sulphide, thiosulphate, ethanol, butyrate, acetate, NADH and hydrogen. Hydrogen acceptors are NAD(P), nitrate, nitrite, possibly dehydrogenase reactions other than via NAD, and, in addition, hydrogen can be released as hydrogen gas. Table 4.1 indicates the distribution of the donors with the systematic arrangement of the photosynthetic bacteria. It appears that whether a bacterium growing by photosynthesis takes up or gives out carbon

Table 4.1 Hydrogen donors in bacterial photosynthesis

Group	Metabolism	Donors	Reductive cycles citrate	Pentose
Green Sulphur Bacteria Chlorobacteriaceae)				
Chlorobium	autotrophic	inorganic S-cpd	+	+
Chloropseudomonas				
ethylicum	myxotrophic	organic-C_2 (CO_2 required)	+	−
Purple Sulphur Bacteria (Thiorhodaceae)				
Chromatium	autotrophic photohetero- trophic	S^{2-}, H_2 organic	−	+
Purple Non-sulphur Bacteria (Athiorhodaceae)				
Rhodomicrobium	photohetero- trophic	organic	−	+
Rhodospirillum *Rhodopseudomonas*	autotrophic photohetero- trophic	H_2, (some)S^{2-} organic	+	+

Note (1) In the heterotrophic cases CO_2 may be either evolved or consumed (assimilated) depending on the particular substrate and the metabolic needs of the cells.
(2) *Rhodospirillum* and *Rhodopseudomonas*, when living chemotrophically, possess an (aerobic) oxidative citrate cycle.
(3) The facultative aerobic phototrophic bacterium *Halobacterium* does not have any clear relation to the above (see Chapter 9). It has heterotrophic metabolism, with no donors.

dioxide depends to a large extent on the hydrogen and carbon sources in the environment. This is the justification for describing nutrition in terms of chemotrophy and phototropy, rather than heterotrophy and autotrophy.

4.56 Energy considerations

According to the scheme of Figure 4.10, about 0·8 eV of the 1·8 eV available per quantum is trapped as oxidoreduction energy. Overall, however, the total energy stored by the non-cyclic electron transport system of green plants is represented by the standard potential difference between oxygen and NADPH, equal to 1·2 eV per electron. Half a molecule of ATP per electron adds 0·17 eV,* so that the energy fixed out of two quanta (3·6 eV) amounts to approximately 1·4 eV. This represents a high efficiency of energy conversion (39%).

* This figure is calculated from the formula $\Delta G_0 = nFE$ taking ΔG_0 as 33 000 J mole^{-1} and $n = 2$. Since the phosphorylation of ADP can proceed well past the standard condition, the actual free energy change, and hence the equivalent calculated redox potential difference may well be twice the above figure.

4.6 The formation of ATP

To approach the problem of the 'coupling' between electron transport and the formation of ATP, it is convenient to consider oxidative phosphorylation in the mitochondrion and photophosphorylation in the chloroplast as the same process in different locations; the problem and the method of attack are the same in each case. The organelle can be treated as a 'black box'; all conceivable models are set out, and their predicted properties compared with the behaviour of the black box under the various available experimental configurations. This approach does not

Electron source:
NADH in mitochondria
Chl* in chloroplasts

$e^- \longrightarrow$

electron acceptor

$\longrightarrow e^-$

$C_{red.} \longrightarrow C_{red.}\!-\!I$

$C_{ox.} \qquad C_{ox.} \sim I$

$X \sim I$

$I \overset{P_i}{\longleftarrow}$

$X \sim P$

$ADP \searrow X$

ATP

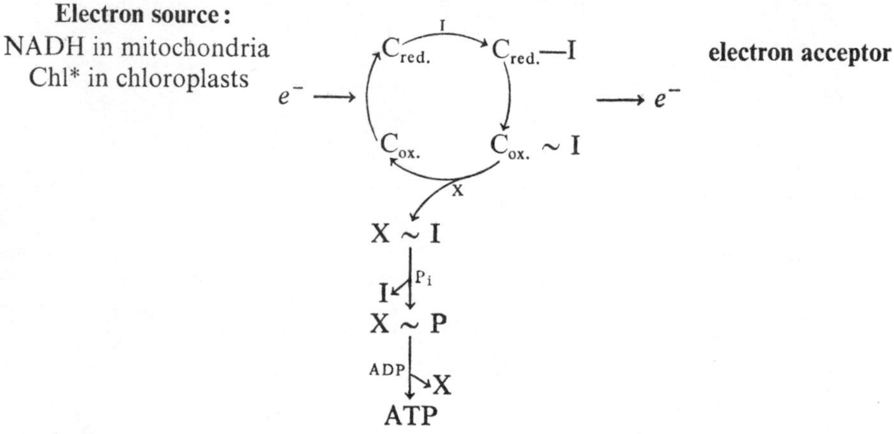

Figure 4.11. Representation of the 'chemical intermediate' hypothesis for the coupling of ATP formation (phosphorylation) to electron transport in either mitochondria or photosynthetic systems

ignore the possibility that different solutions may be found in the end for the two organelles.

Two kinds of model are prominent, and it is not possible at present to recommend either before the other. The older model, the chemical intermediate hypothesis, envisages that one of the redox materials at a given phosphorylation site forms a chemical bond with a high negative free energy of hydrolysis, which leads by transfer reactions with energy conservation to 'high-energy' phosphorylated compounds which transfer their phosphate to ADP in a kinase reaction. There is in this view a 'transferase chain' in addition to the oxidoreductase chain of electron transport. Components of such a chain could be called coupling factors, without which ATP formation could not occur. 'Uncoupling' which is observed as unrestricted electron transport with no phosphorylation taking place, would be interpreted as a hydrolytic breakdown of one of the substrates of the transferase chain (see Figure 4.11).

The alternative, newer hypothesis is that put forward by Mitchell, and termed the chemiosmotic hypothesis. In this view electron carriers are so located in the membrane material that hydrogen ions are pumped from one face to the other.

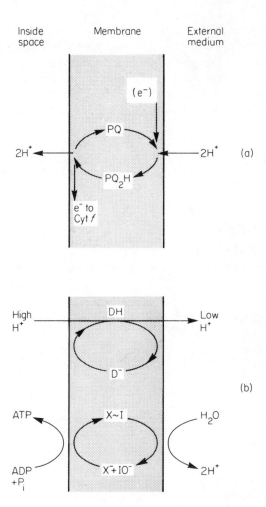

Figure 4.12. The 'chemiosmotic' hypothesis for coupling phosphorylation to electron transport. In the upper diagram electron transport causes an accumulation of H⁺ ions on one side of a membrane, which may exchange with say K⁺ to produce an electric field, the combined effect being to operate the system shown in the lower diagram, which may be thought of as an ATP-driven H⁺-pump running in reverse

(One should remember that the membranes of mitochondrial cristae and chloroplast thylakoids where electron transport and phosphorylation occur are seen to be much thicker than other cell membranes in sections under the electron microscope.) This is brought about by the alternation of carriers such as cytochromes, which carry electrons, and those such as flavins, which carry hydrogen atoms. Hydrogen ions are presumed to be taken up at one face by the reduction of the second by the first type, and released at the other face by the reduction of a further member of the first type by the second (see Figure 4.12).

The model continues by postulating substances that condense on the acid side, migrate across the membrane (conveying the hydrogen ions down the gradient) and then dissociate again by phosphorylase and kinase reactions. The coupling factors in this case are the systems carrying out the final process. Uncoupling on this theory is a process which interferes with the integrity of the membrane, or allows recombination of the separated ions other than through the coupling site, or (in the case of ammonium ions and primary amines) neutralizes the hydrogen-ion gradient.

On both hypotheses, the final enzymes should have ATPase activity, and particles are known (see Plate 9) from the membrane surface of mitochondria, chloroplasts and bacteria which have such activity, and have the properties of coupling factors.

4.7 van Niel's Hypothesis

A major step in the comparison of photosynthetic systems was made by van Niel in 1933. He rationalized the great variety of chemical reactions carried out

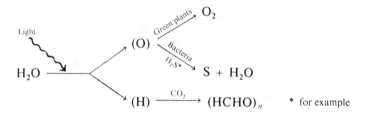

Figure 4.13 Van Niel's hypothesis, that in all photosynthetic processes water is split to give an oxidant and a reductant

in bacterial photosynthesis and in the green-plant process involving oxygen production by supposing that the initial step of photosynthesis was the splitting of water. This photolysis of water led to the formation of an oxidant and a reductant. The difference between plants and bacteria on this basis was that green plants had the means to release oxygen from the oxidant as gas, while the anaerobic bacteria released it by combination with an environmental reductant (see van Niel, 1935). This hypothesis, represented by the diagram in Figure 4.13, stimulated research for some thirty years. However the elucidation of electron transport chains has shown now that separation of charge explains the same observations and is easier to understand as the primary photochemical process. When as must usually happen the elements of water enter into the half-cell reactions at either end of the electron transport pathway, this can be regarded as an ionic reaction with either hydrogen or hydroxyl ions. The formation of these is a fast spontaneous reaction, and in no sense a photolysis. Nevertheless, the simultaneous removal of hydrogen and hydroxyl ions by the two electron transport termini requires the same energy and has the same overall formulation as the hypothetical photolysis of a water molecule. A modern translation of van

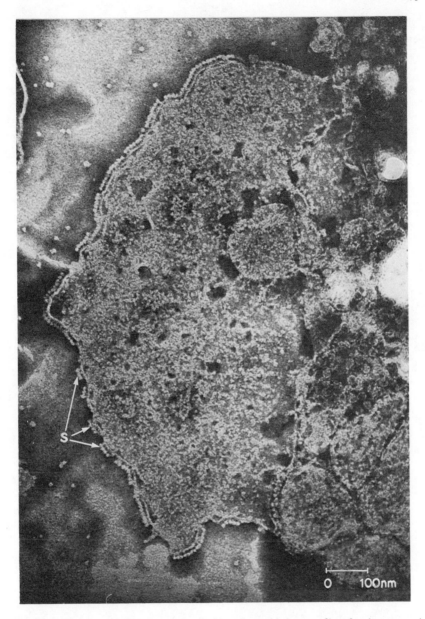

Plate 9. Fragmented thylakoids from spinach chloroplasts, dried onto a film of carbon-covered collodion and negatively contrasted by deposition of phosphotungstic acid (adjusted to pH 7). The 9 nm stalked spheres (S) are clearly visible at the edge of the large fragment. By courtesy of M. Raymond Bronchart, Department de Botanique, Universite de Liège, From C. Sivonval, (ed.) *Le Chloroplaste,* Masson, Paris, 1967, p. 55, plate IX(B), with permission of the publisher.

Niel's hypothesis could be that photosynthesis involves a *separation of charge* leading to the formation of an oxidant and a reductant.

4.8 Summary

The thylakoids of the green plant in conjunction with ferredoxin evolve oxygen and reduce NADP in light, and a widely accepted possible mechanism is represented in Figure 4.10. By this means ATP and NADPH are supplied to the stroma enzymes, probably in equimolar proportions. This is the non-cyclic process of electron transport and photophosphorylation.

In the cyclic process (Figure 4.7) chloroplast fragments supplied with ferredoxin form ATP in the light. By the two processes together the thylakoids can supply ATP in a higher proportion to NADPH.

Both processes probably occur in the bacteria.

Suggested further reading

Hill, R. (1965). The Biochemists' Green Mansions: the photosynthetic electron-transport chain in plants. In P. N. Campbell and G. D. Greville (Eds.), *Essays in Biochemistry*, Vol. 1, Academic Press, London, p. 121.

Arnon, D. I. (1968). Electron Transport and Photophosphorylation in Chloroplasts. In T. P. Singer (Ed.), *Biological Oxidations*, Interscience, New York, p. 123.

Glasstone, S. and D. Lewis (1960). *Elements of Physical Chemistry*, 2nd ed., Chapter 13, Macmillan, London.

Hall, D. O. and K. K. Rao (1972). *Photosynthesis*, Arnold, London.

Clayton, R. K. (1971). *Light and Living Matter*, Vol. 2, McGraw-Hill, New York.

CHAPTER 5

The path of carbon

The model so far presented separates photosynthesis into 'light' and 'dark' reactions, the light reactions being the pigment-driven electron transport pathways (Chapter 4) which result in the formation of NADPH and ATP. The dark reactions now to be described involve a series of enzymes in the stroma of the chloroplast which use the ATP and NADPH for the reduction of carbon dioxide, so that ATP and NADPH are the only link between the thylakoid and stroma. Similar relationships are supposed to hold between both the chromatophores of photosynthetic bacteria and the thylakoids of blue–green algae and their respective cytoplasms. While there are still some difficulties in this model, it has great strength since there can be no doubt that thylakoid preparations can be made to produce ATP and NADPH, and also that enzymes do exist in the stroma which given these materials will reduce carbon dioxide.

In this section we shall consider sequences of enzyme-catalysed reactions that have been put together as 'metabolic pathways', to explain the formation of cell material from the carbon sources used by photosynthetic organisms. In the green plants carbon dioxide is usually used to make carbohydrate. In some cases other products may predominate, such as fat or glycollate derivatives; this is particulary important in algae. In green plants, however, it appears to be the case that all carbon dioxide fixation proceeds via one pathway, the *reductive pentose cycle*, and the essential features of this pathway will be set out in detail. This cycle is supported in some cases by interesting pathways of which two spectacular examples are the *Crassulacean acid metabolism* (CAM) and the C_4-*dicarboxylic acid pathway*, also known as the Hatch–Slack pathway. These serve to accumulate carbon dioxide under conditions which cause CO_2-limitation in 'normal' (C_3) plants. Their outlines and importance are discussed in Chapter 10. Photosynthetic bacteria, on the other hand possess two (at least) entirely separate pathways for carbon dioxide fixation; although most species operate the reductive pentose cycle, a reductive citrate cycle is known. Many species appear to grow fastest when carbon is supplied in an organic form such as acetate. These three pathways of carbon uptake will be compared in this chapter.

5.1 The incorporation of carbon dioxide in the chloroplast

This pathway for the incorporation of carbon dioxide and its reduction to carbohydrate was discovered by the group at Berkeley led by Calvin. It is cyclic in operation, and has been variously termed the 'Calvin Cycle', the 'Calvin–Benson–Bassham Cycle', and the 'Photosynthetic Carbon Cycle'. However,

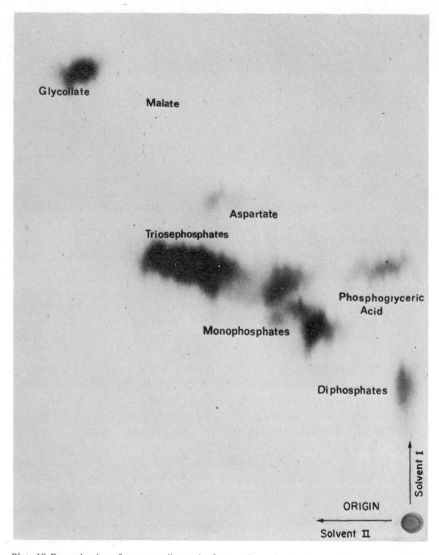

Plate 10. Reproduction of an autoradiograph of a two-dimensional chromatogram on paper of the products of photosynthesis in pea chloroplasts, using [^{14}C] CO_2. The dark spot due to radioactive triosephosphate indicates that the principal pathway was the reductive pentose cycle, other intermediates of which can be seen as paler spots. The occurrence of glycollate is common, and is discussed in Chapter 10. Autoradiogram by courtesy of Professor D. A. Walker, Department of Botany, The University, Sheffield.

although it may well represent the principal pathway for photosynthesis in most green plants under 'normal conditions' ('normal conditions' referring to open-air, temperate-zone vegetation in a maritime climate in sunlight!) other cyclic pathways do exist in bacteria and some green plants. We will use the term 'Reductive Pentose Cycle' for the sequence of reactions set out in Figure 5.1.

The cycle first appeared in a paper by Bassham and coworkers (1954) and it was based primarily on the use of the newly available isotope ^{14}C. Cultures of *Chlorella* were allowed to incorporate $[^{14}C]\text{-}CO_2$ for limited periods of time. The cells were then killed and their contents separated and analysed by the relatively new technique of *chromatography*, usually in two dimension on paper. Each metabolic intermediate was located at a particular point on the paper in the form of a 'spot', and was detected either by chemical reagents which produced a colour reaction, or by *radioautography*. (In the latter method an X-ray film is clamped against the chromatogram, and developed after an exposure varying from minutes to days. The intensity of the dark areas indicates the quantity of the radioactive isotope.) Thus it was found that the first material to be labelled was 3-phosphoglyceric acid (PGA), and the other intermediates of the cycle shown in Figure 5.1 (with the exception of sedoheptulose-1,7-bisphosphate*) appeared at later times. The compounds were analysed, so that the distribution of the labelled isotope among the carbon atoms of the intermediates was established. It was found that the first label was in the 1-position (carboxyl), but that at longer times it appeared in carbons 2 and 3. This meant that the acceptor for CO_2 was formed from the product, and hence a cycle was indicated. Plate 10 is a typical radioautograph of an experiment of this type. The above workers and also the group led by Arnon (also at Berkeley) showed that the algae contained enzymes that would catalyse the reactions that had been postulated. The elucidation of the pathway was a great achievement, recognized by the award of a Nobel prize to Calvin in 1961.

Figure 5.1 sets out the reactions of the reductive pentose cycle so as to show the overall formation of carbohydrate in the formation of D-glyceraldehyde-3-phosphate from 3 molecules of carbon dioxide. Glyceraldehyde-3-phosphate (G3P) is a phosphate ester of a *triose*: trioses are monosaccharide sugars of three

$$
\begin{array}{l}
\text{CHO} \\
| \\
\text{HCOH} \\
| \\
\text{CH}_2\text{O}\textcircled{P}
\end{array}
\qquad
\textcircled{P} \text{ represents } \text{—PO(OH)}_2 \\
\qquad\qquad\quad\text{(orthophosphate)}
$$

G3P

---

* The term 'bisphosphate' replaces the older term 'diphosphate' in the following cases: fructose-1,6-bisphosphate, FBP (fructosediphosphate, FDP); 1,3-bisphosphoglyceric acid, BPGA (diphosphoglyceric acid, DPGA); ribulose-1,5-bisphosphate, RuBP (ribulosediphosphate, RuDP); sedoheptulose-1,7-bisphosphate, SHBP (sedoheptulosediphosphate, SHDP); ribulosebisphosphate carboxylase, RuBP carboxylase, EC.4.1.1.39 (RuDP carboxylase).

Although this usage will be found in very few works published before 1976 it is necessary to abide by it in this text, and it is hoped that the student will make the necessary translation when referring to the established literature.

78

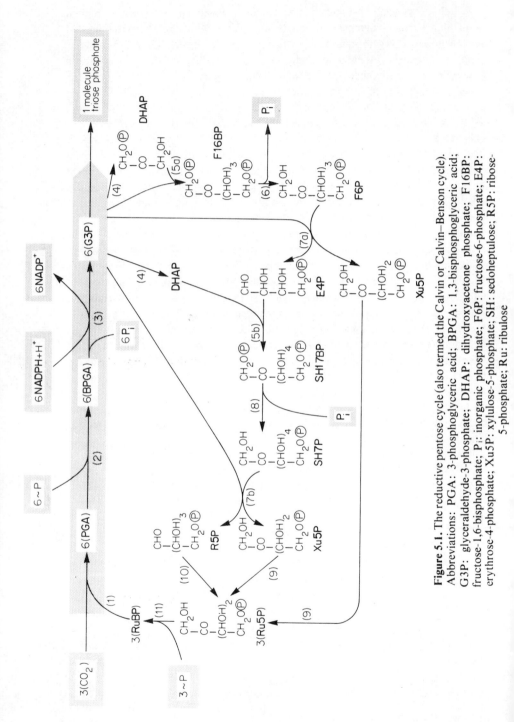

Figure 5.1. The reductive pentose cycle (also termed the Calvin or Calvin–Benson cycle). Abbreviations: PGA: 3-phosphoglyceric acid; BPGA: 1,3-bisphosphoglyceric acid; G3P: glyceraldehyde-3-phosphate; DHAP: dihydroxyacetone phosphate; F16BP: fructose-1,6-bisphosphate; F6P: fructose-6-phosphate; E4P: erythrose-4-phosphate; P_i: inorganic phosphate; Xu5P: xylulose-5-phosphate; SH: sedoheptulose; R5P: ribose-5-phosphate; Ru: ribulose

carbon atoms. The carbohydrates involved in the reductive pentose cycle include tetroses (4 carbons), pentoses (5), hexoses (6) and heptoses (7). Although G3P is an important biochemical material, it is not present in large quantities, and the product of photosynthesis that is usually observed is sucrose (made of one molecule each of the hexoses glucose and fructose) or starch (a large molecule formed by condensation of many glucose units). The interconversion of carbohydrates is, however, a metabolic process occuring in most organisms, including the classical materials of biochemical study: yeast cells, muscle and liver tissues. Once carbohydrate has been fed into this system in almost any form it can be easily converted to any other form required by the cell at the time. It is therefore convenient to regard the specifically photosynthetic part of the process as terminated with the production of carbohydrate, here triose phosphate.

The reactions of the cycle are identified both by name and, in the text, by their reference numbers in the Enzyme Commission scheme of classification. The main features of this scheme are as follows. There are four numbers in a reference, the first denoting which of six classes of reaction the enzyme catalyses. In class EC.1, the *oxidoreductases* catalyse reactions of the type

$$AH_2 + B \xrightarrow{\text{EC.1}} A + BH_2$$

In the cycle, an example is the reduction of diphosphoglyceric acid (DPGA) to G3P by the enzyme EC1.2.1.13. The second number here indicates that the substrate *acid* is reduced to *aldehyde* and the third that the hydrogen is carried by a nicotinamide coenzyme, in this case NADP:

$$NADPH + \begin{array}{c} COO\,\textcircled{P} \\ | \\ HCOH \\ | \\ CH_2O\,\textcircled{P} \end{array} \xrightarrow{\text{EC.1.2.1.13}} \begin{array}{c} CHO \\ | \\ HCOH \\ | \\ CH_2O\,\textcircled{P} \end{array} + NADP^+ + H^+ + P_i$$

The fourth number is a specic serial number.

The second class, of transferring enzymes (*transferases*) may be represented by the scheme:

$$AB + C \xrightarrow{\text{EC.2}} A + BC$$

The second and third numbers describe the nature of the group transferred ('B') and sometimes the nature of the acceptor ('C'). The fourth number as always is the specific serial reference.

The third type of reaction is *hydrolysis* which is the breaking of a molecule by the addition of the elements of water, thus:

$$AB + H_2O \xrightarrow{\text{EC.3}} AH + BOH$$

Enzymes of this group are the *hydrolases*, and in the reductive pentose cycle they are represented by *phosphatases*, where the group 'B' in the above equation is the phosphate group that we have shown by the symbol $\textcircled{P}$.

Class EC.4, the lyases, nominally break molecules into two products:

$$AB \rightarrow A + B$$

but there are two points to make. First, the reaction may be found going the opposite way from the nominal direction above, and in the examples of the *aldolase* enzymes of the cycle, the reactions are indeed reversible. Secondly, the key enzyme of the cycle that adds carbon dioxide to a pentose, producing two molecules of PGA, is classified as a lyase, which might be thought surprising.

Class EC.5 contains the *isomerases*, which interconvert pairs of chemical isomers:

$$ABC \xrightarrow{\text{EC.5}} ACB$$

We may distinguish *epimerases* (EC.5.1) which invert the stereochemical configuration at a specific point, from the *intramolecular oxidereductases* (EC.5.3) which here interconvert aldose and ketose sugars:

$$
\begin{array}{ccc}
\begin{array}{l} CHO \\ | \\ HCOH \\ | \\ CH_2O\,\textcircled{P} \end{array}
&
\xrightleftharpoons{\text{EC.5.3.1.1}}
&
\begin{array}{l} CH_2OH \\ | \\ CO \\ | \\ CH_2O\,\textcircled{P} \end{array}
\end{array}
$$

G3P
(aldotriose
phosphate)

Dihydroxyacetone
phosphate
(ketotriose phosphate)

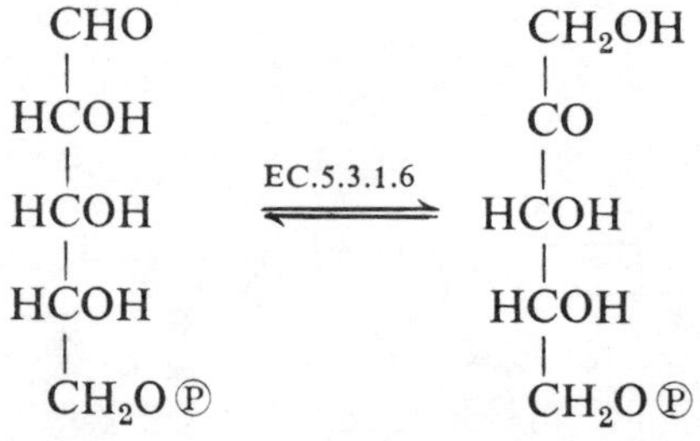

$$
\begin{array}{ccc}
\begin{array}{l} CHO \\ | \\ HCOH \\ | \\ HCOH \\ | \\ HCOH \\ | \\ CH_2O\,\textcircled{P} \end{array}
&
\xrightleftharpoons{\text{EC.5.3.1.6}}
&
\begin{array}{l} CH_2OH \\ | \\ CO \\ | \\ HCOH \\ | \\ HCOH \\ | \\ CH_2O\,\textcircled{P} \end{array}
\end{array}
$$

Ribose-5-phosphate
(aldopentose
phosphate)

Ribulose-5-phosphate
(ketopentose phosphate)

In general, ketosugars are named after a parent aldosugar, inserting the letters -ul-, thus ribulose is the ketosugar related to ribose (and also arabinose). An example of an epimerase reaction is the converstion of xylulose to ribulose phosphate by epimerization at the 3-position:

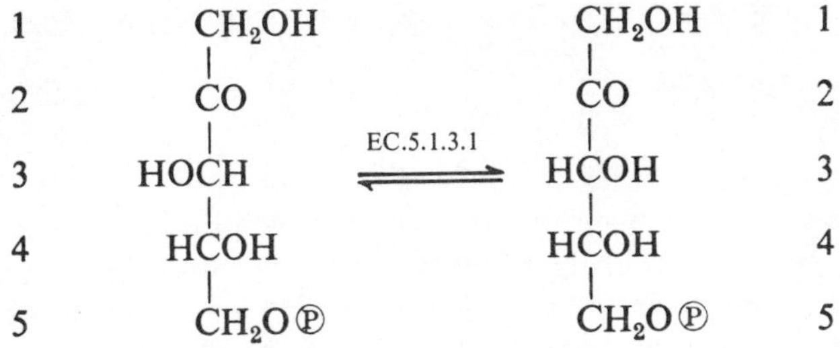

$$
\begin{array}{ccc}
1 & CH_2OH & CH_2OH & 1 \\
& | & | & \\
2 & CO & CO & 2 \\
& | & \xrightarrow{EC.5.1.3.1} & | & \\
3 & HOCH & \rightleftharpoons & HCOH & 3 \\
& | & | & \\
4 & HCOH & HCOH & 4 \\
& | & | & \\
5 & CH_2O\,\text{\textcircled{P}} & CH_2O\,\text{\textcircled{P}} & 5 \\
\end{array}
$$

The sixth class, the ligases, represented by the general equation

$$A + B + ATP \xrightarrow{EC.6} AB + ADP + P_i$$

are not found in the reductive pentose cycle.

5.11 Description of the reductive pentose cycle

Referring to Figure 5.1, we shall commence with the entry of carbon dioxide: the enzyme *ribulosebisphosphate carboxylase*, also known as carboxydismutase, EC.4.1.1.39, carries out the reaction (1):

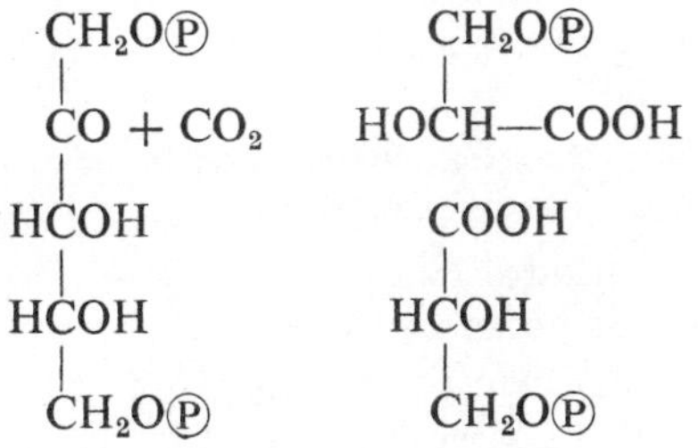

$$
\begin{array}{cc}
CH_2O\,\text{\textcircled{P}} & CH_2O\,\text{\textcircled{P}} \\
| & | \\
CO + CO_2 & HOCH-COOH \\
| & \\
HCOH & COOH \\
| & | \\
HCOH & HCOH \\
| & | \\
CH_2O\,\text{\textcircled{P}} & CH_2O\,\text{\textcircled{P}} \\
\end{array}
$$

$$RuBP + CO_2 \xrightarrow{Mg^{2+}} 2PGA \tag{1}$$

(RuBP stands for ribulosebisphosphate)

In this reaction one would suspect a six-carbon intermediate, but in spite of much searching, none has been found. The activity of the enzyme as isolated has been recently improved by the development of better procedures, better assay methods and the presence of allosteric activators. These points, plus the surprising reaction with oxygen, are discussed in section 10.21. The enzyme constitutes nearly half of all leaf protein and is often referred to as 'Fraction-I' protein.

The second step (2) (from our arbitrary starting point) is one of the two sites where ATP is required; ATP is a necessary source of energy, and is provided by the process of photophosphorylation in the thylakoids (section 4.6). Here the terminal phosphate group of ATP is transferred to the carboxyl group (carbon-1) of

$$\begin{array}{c} \text{COOH} \\ | \\ \text{HCOH} \\ | \\ \text{CH}_2\text{O}\textcircled{P} \\ \text{PGA} \end{array} + \text{ATP} \underset{\substack{\text{phosphoglycerate kinase} \\ \text{EC.2.7.1.3}}}{\rightleftharpoons} \text{ADP} + \begin{array}{c} \text{CO.O}\textcircled{P} \\ | \\ \text{HCOH} \\ | \\ \text{CH}_2\text{O}\textcircled{P} \\ \text{1,3-bisphosphoglyceric} \\ \text{acid (1,3-BPGA)} \end{array} \quad (2)$$

PGA to make a 'mixed anhydride'. As in the case of the glycolysis pathway where the same enzyme acts, the equilibrium is in favour of the synthesis of ATP. However the impetus of the cycle is maintained by the overall energy change, which is large enough to make the cycle as a whole effectively irreversible.

1,3-Bisphosphoglyceric acid (1,3BPGA) is a substrate for the enzyme *triosephosphate dehydrogenase* (or more formally D-glyceraldehyde-3-phosphate dehydrogenase) EC.1.2.1.13 acting in the reverse direction to its name. This step (3) provides the site of entry for the third and last 'raw material' of the cycle: the reduced coenzyme NADPH. It may be noted that the corresponding enzyme in the glycolysis pathway, EC.1.2.1.12, uses the coenzyme NAD. In a plant cell the former occurs inside the chloroplast, the latter outside in the cell cytoplasm. The NADP-coupled enzyme is much more difficult to isolate. The product of the reaction (3), glyceraldehyde-3-phosphate, is the final product that we have set ourselves to produce; however, we have only accounted for the uptake of one molecule of carbon dioxide, and have not explained the origin of the ribulosebisphosphate used in the first step. The rest of the cycle therefore will take five-sixths of the triosephosphate and reconvert it to ribulose-1,5-bisphosphate. This process will consume the remaining requirement of ATP. The reactions of this regenerative part of the cycle are almost exactly those of the pentose phosphate pathway in animals and bacteria, which is why it is instructive to refer to the overall process as the 'reductive' pentose cycle.

The reactions catalysed by *triosephosphate isomerase* EC.5.3.1.1, *aldolase* EC.4.1.2.13 and fructose *bisphosphatase* EC.3.1.3.11 (4, 5 and 6 respectively) are those of glucogenesis; transketolase EC.2.2.1.1. is apparently the same as in the pentose phosphate pathway of animals. Transketolase occurs twice in Figure 5.1, in the reactions (7a) and (7b).

$$\begin{array}{c} \text{CH}_2\text{OH} \\ | \\ \text{CO} \\ | \\ \text{HOCH} \\ | \\ \text{HCOH} \\ | \\ \text{HCOH} \\ | \\ \text{CH}_2\textcircled{P} \end{array} + \begin{array}{c} \text{CHO} \\ | \\ \text{HCOH} \\ | \\ \text{CH}_2\text{O}\textcircled{P} \end{array} \underset{\text{Mg}^{2+}}{\overset{\text{TPP}}{\rightleftharpoons}} \begin{array}{c} \text{CHO} \\ | \\ \text{HCOH} \\ | \\ \text{HCOH} \\ | \\ \text{CH}_2\text{O}\textcircled{P} \end{array} + \begin{array}{c} \text{CH}_2\text{OH} \\ | \\ \text{CO} \\ | \\ \text{HOCH} \\ | \\ \text{HCOH} \\ | \\ \text{CH}_2\text{O}\textcircled{P} \end{array} \quad (7a)$$

Fructose-6 phosphate (F6P)   G3P   Erythrose-4-phosphate (E4P)   Xylulose-5-phosphate (Xu5P)

$$
\begin{array}{ccc}
\begin{array}{l}
CH_2OH \\
| \\
CO \\
| \\
HOCH \\
| \\
HCOH \\
| \\
HCOH \\
| \\
HCOH \\
| \\
CH_2O\textcircled{P}
\end{array}
\;+\;
\begin{array}{l}
CHO \\
| \\
HCOH \\
| \\
CH_2O\textcircled{P}
\end{array}
\;\underset{Mg^{2+}}{\overset{TPP}{\rightleftharpoons}}\;
\begin{array}{l}
CHO \\
| \\
HCOH \\
| \\
HCOH \\
| \\
HCOH \\
| \\
CH_2O\textcircled{P}
\end{array}
\;+\;
\begin{array}{l}
CH_2OH \\
| \\
CO \\
| \\
HOCH \\
| \\
HCOH \\
| \\
CH_2O\textcircled{P}
\end{array}
& (7b)
\end{array}
$$

Sedoheptulose-7   G3P   Ribose-5-   X5P
phosphate   phosphate
(SH7P)   (R5P)

It transfers the carbons 1 and 2 (as a unit) from a ketose to an aldose acceptor; since the remains of the donor is now an aldose (with 2 carbons less) and the acceptor has become a ketose (with 2 carbons more) the reaction is reversible.

Thiamine pyrophosphate acts as the C_2-carrier, and it appears that the pool of TPP-glycolaldehyde adduct is common to the two reactions. *Glycollate* formed under some conditions may originate from this pool: this will be discussed later (Chapter 10).

The enzyme *aldolase*, EC.4.1.2.13 which adds dihydroxyacetone phosphate (DHAP) to an aldehyde thus forming a ketose-1-phosphate, also has two sites of action, (5a) where the aldehyde is G3P, making fructose-1,6-bisphosphate (F16DP), and another (5b), that may possibly be a different form of the enzyme, that makes sedoheptulose-1,7-bisphosphate (SH17DP) using erythrose-4-phosphate (E4P):

$$
\begin{array}{ccc}
\begin{array}{l}
CH_2O\textcircled{P} \\
| \\
CO \\
| \\
CH_2OH
\end{array}
\;+\;
\begin{array}{l}
CHO \\
| \\
HCOH \\
| \\
HCOH \\
| \\
CH_2O\textcircled{P}
\end{array}
\;\overset{Aldolase}{\longrightarrow}\;
\begin{array}{l}
CH_2O\textcircled{P} \\
| \\
CO \\
| \\
HOCH \\
| \\
HCOH \\
| \\
HCOH \\
| \\
HCOH \\
| \\
CH_2O\textcircled{P}
\end{array}
& (5b)
\end{array}
$$

DHAP   E4P   SH17BP

There is a specific hydrolase, sedoheptulose bisphosphatase, that removes the 1-phosphate, leaving SH7P (8).

The pentose phosphates Xu5P and R5P are both converted (9, 10) to Ru5P by means of the enzymes ribulosephosphate-3-epimerase EC.5.1.3.1 and ribosephosphate isomerase EC.5.3.1.6 respectively. Finally the ribulose-5-phosphate is phosphorylated (11) at the 1-position by ATP (completing the account of the 'raw materials') giving RuBP and completing the cycle.

In summary, the reductive pentose cycle has been described as the uptake of three molecules of carbon dioxide by three molecules of RuBP giving six PGA, of which one is a product; the remaining five PGA molecules (totalling fifteen carbon atoms) are rearranged into three molecules of RuBP. During this cycle nine molecules of ATP and six of NADPH are consumed. This can be represented in one line:

$$3RuBP + 3CO_2 \xrightarrow{9ATP + 6NADPH} G3P + 3RuBP$$

The pentose phosphate pathway in animals involves the enzyme *transaldolase*, avoiding the formation of sedoheptulose-1,7-bisphosphate. However, since transaldolase has been shown to exist only in a few plant species, and in these only in small quantities, it has been concluded that it is not involved in the reactions of the reductive pentose phosphate cycle.

Although many of the enzymes are common to pathways in animal tissues, studies in the latter indicate an elaborate series of controls, which may differ from source to source. The classification of chloroplast enzymes into general categories may conceal important differences. The discovery of allosteric and other regulatory effects in the enzymes of, say, glycolysis has led to some remarkable conclusions, and it can be expected that enzymes of the above cycle will be found to possess a similar degree of control. Some of the possible control mechanisms are discussed in Chapter 10.

It was not until about ten years after the appearance of the reductive pentose cycle that techniques of isolating chloroplasts reached the point where it could be said that all the enzymes of the pathway were contained inside the envelope, and in sufficient concentration to enable the cycle to operate at the expected rate. If chloroplasts are prepared quickly (after the leaves have been kept in the dark to remove most of the heavy starch grains) they will when illuminated take up carbon dioxide and evolve oxygen. Carbon dioxide is the ultimate, natural, Hill reagent.

5.2 Carbon metabolism in the phototrophic bacteria

Chromatophores from all three main groups of photosynthetic bacteria carry out a cyclic photophosphorylation process. Non-cyclic electron transport has been observed, ending with the reduction of ferredoxin, or NAD (not apparently NADP); the reduction of NAD is independent of ferredoxin. The question whether this non-cyclic process is directly light-driven can be left until a later chapter; for the time being, it should be remembered that the membrane material of photosynthetic bacteria when illuminated forms NADH and ATP.

5.21 Incorporation of organic materials

While some bacteria can photosynthetically reduce carbon dioxide with an inorganic hydrogen donor, others are able to use organic materials as sources of carbon and hydrogen. The second case will be considered first. All three groups of bacteria carry out such a process, although the biochemical means differ. There are two separate dark reactions, one which oxidizes the substrate, producing reducing equivalents, and the other reductive incorporating the substrate into cell constituents. This kind of process is known in chemistry as a 'disproportionation' reaction, (Figure 5.2), and can occur spontaneously in a few

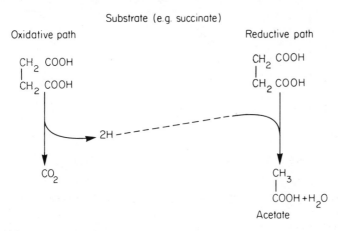

Figure 5.2. An oxidative and a reductive pathway coupled to form a 'disproportionation' reaction

cases. In a photosynthetic context the substrates available do not undergo the reactions spontaneously (which is the same as saying that they are close to a state of chemical equilibrium), and energy must be supplied to the system to cause it to run in the direction indicated. This energy of course comes from light, and may be mediated in two ways. First, the excitation energy from the light may be used to drive a non-cyclic electron transport system, which takes electrons at one potential, and drives them to a much lower (more reducing) potential (Figure 5.3). (It should be stressed that electrons are not available from water, in contrast to non-cyclic electron flow in green plants.) There is some doubt whether non-cyclic, light driven electron flow actually takes place in bacteria; the same result could be obtained by the second mediation system, using ATP provided by cyclic, light-driven phosphorylation. By coupling a reaction to the splitting of ATP, the total change in free energy may be sufficiently negative to carry it on. ATP could be applied to the system of Figure 5.2 at two possible sites (Figure 5.4).

The green sulphur bacterium *Chloropseudomonas ethylicum* oxidizes acetate by the citric acid cycle (an outline of this is shown in Figure 5.5), and it possesses the modification to this pathway known as the glyoxylate pathway which

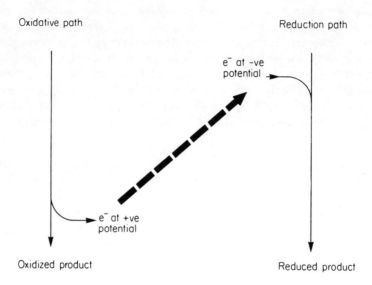

Figure 5.3. The system of Fig. 5.2 driven by light energy

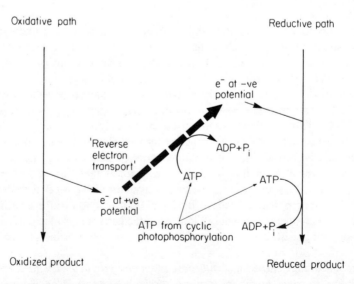

Figure 5.4. The system of Fig. 5.3 driven by ATP from either cyclic photophosphorylation or elsewhere

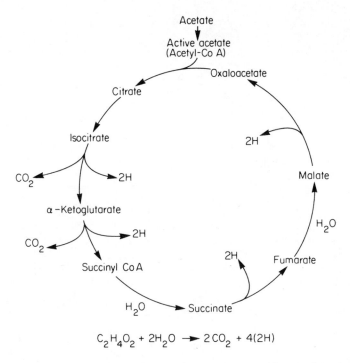

Figure 5.5. The citrate cycle common to most aerobic organisms

enables it to metabolize acetate avoiding the immediate loss of two carbon atoms as carbon dioxide (see Figure 5.6). The reducing equivalents available from the oxidation path via the citric acid cycle, with ATP, can be used to reduce the organic acids shown to materials such as fats, carbohydrates and aminoacids.

5.22 Incorporation of carbon dioxide in photosynthetic bacteria

The purple sulphur bacterium *Chromatium* can be grown on acetate in the same way, and probably using the same metabolic pathways as *Chloropseudomonas* above. This organism is however able to grow autotrophically, reducing carbon dioxide by means of an inorganic hydrogen donor such as hydrogen sulphide. Under these circumstances the bacterial cells cease to synthesize the enzymes of the glyoxylate pathway, and begin to make those of the reductive pentose pathway (section 5.1). The switching of metabolism from one route to another by means of the selective synthesis of the necessary enzymes is a common feature in the bacteria, and must account to a large extent for their great ability to survive changes in conditions.

Another species of green sulphur bacterium *Chlorobium thiosulphatophilum* possesses both the reductive pentose cycle and a new process recently discovered in the laboratory of Arnon and termed the *reductive citric acid cycle*. This process reverses the direction of the citric acid cycle of Figure 5.5 by means of

88

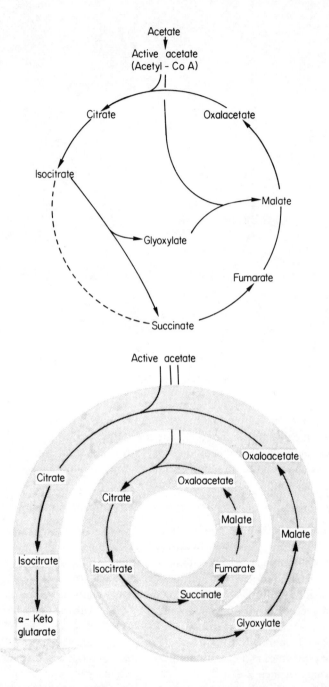

Figure 5.6. The glyoxylate pathway drawn to show its relation to the citrate cycle (a), and (b) drawn to show its operation as a cycle for synthesis

four enzymes which are adapted to reverse stages, that are normally virtually irreversible, by coupling to them energy in the form of ATP, or as the strongly-reducing agent, reduced ferredoxin. This reductive citrate cycle is shown in Figure 5.7. The key enzymes are, first, the reversal of the condensation of active

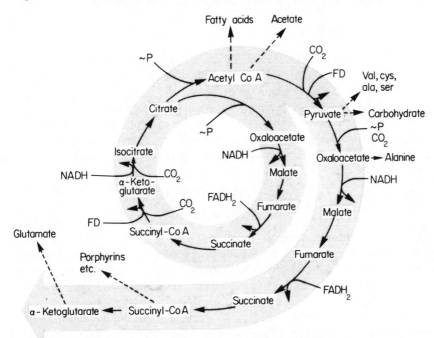

Figure 5.7. The reductive citrate cycle drawn to show its biosynthetic value. After Evans and coworkers (1966). *Proc. Natl. Acad. Sci.*, **55**, 928, with permission

acetate ('acetyl CoA') with oxaloacetate to give citrate; the reversal splits ATP to ADP and inorganic phosphate:

$$\text{Citrate} + \text{CoA} + \text{ATP} \xrightarrow{\text{EC.4.1.3.8}} \text{Oxaloacetate} + \text{Acetyl CoA} + \text{ADP} + \text{P}_i$$

This enzyme, the 'citrate cleavage enzyme' is known in animal cells, but is located away from the region where the citric acid cycle takes place.

Two enzymes reverse the 'oxidative decarboxylation of α-ketoacids' in which pyruvate and α-ketoglutarate form acetyl and succinyl CoA derivatives respectively, losing carbon dioxide and reducing the coenzyme NAD (Figure 5.7). The reversal is carried out, not with NADH, but with reduced ferredoxin. The extra 100–150 mV of reducing potential is apparently sufficient to allow the reaction to proceed. The fourth enzyme phosphorylates pyruvic acid to phosphoenolpyruvate, obtaining energy by splitting both high-energy bonds of the ATP so as to yield AMP and inorganic phosphate (cf. EC.2.7.9.1, p. 188):

$$
\begin{array}{ccc}
\text{COOH} & & \text{COOH} \\
| & & | \\
\text{CO} & + \text{ATP} \longrightarrow \text{AMP} + \text{P}_i + & \text{CO}\circled{P} \\
| & & \| \\
\text{CH}_3 & & \text{CH}_2
\end{array}
$$

(This is not related to the citric acid cycle as such but rather to a sequence of metabolic paths connecting it with the glycolysis system. In liver tissue, the overall reaction of the conversion of pyruvate to oxaloacetate takes place directly using ATP and the coenzyme biotin.) The bacteria in which this pathway has been demonstrated are all obligate anaerobes, in that they will not grow if even traces of oxygen are present. If this is due to the tendency of strongly reducing materials to be oxidized by oxygen with the production of superoxide ions, or hydrogen peroxide

$$A^- + O_2 \rightarrow A + O_2{}^-,$$

$$AH_2 + O_2 \rightarrow A + H_2O_2$$

then this raises the question, so far unanswered, of how this is avoided in the chloroplast system. (Ferredoxin has also been obtained from non-photosynthetic bacteria such as the (anaerobic) *Clostridia*, which also contain the enzymes for the formation of pyruvate and α-ketoglutarate from acetyl and succinyl CoA.)

The purple non-sulphur bacteria can live either phototrophically or chemotrophically: the latter existence requires oxygen and a respiratory substrate such as acetate, and the photosynthetic pigments are no longer produced. The colour of the cells then fades almost completely as the bacteria grow. When the oxygen supply is exhausted, the cells given light reform their red pigment and the underlying bacteriochlorophyll and re-commence photosynthesis. The photosynthetic incorporation of acetate does not appear to involve the glyoxylate pathway, and may involve the simultaneous fixation of carbon dioxide. These bacteria, like the green and purple sulphur bacteria, contain ferredoxin, and its function in carbon assimilation is not established. However there is a process whereby hydrogen gas can be evolved from substrates such as acetate, succinate, fumarate and L-malate, which are completely oxidized to carbon dioxide. For example

$$CH_3.COOH + 2H_2O \rightarrow 2CO_2 + 4H_2$$

These photodecompositions take place fastest either with glutamate as the nitrogen source, or when the nitrogen source is exhausted. Since the energy yields are small, it is not likely that this is primarily a nutritive process. It might be the case that, as the hydrogenase enzyme is close to the site of nitrogen fixation, the evolution of hydrogen is a means of ensuring maximum activity for the fixation of nitrogen gas from the air. A strong reductant (ferredoxin rather than NADH) is required for nitrogen fixation, and since ferredoxin has a lower potential than the standard hydrogen electrode at pH 7, activation of the nitrogen fixing process would be expected to result in hydrogen gas liberation by the hydrogenase. Both the 'nitrogenase' and 'hydrogenase' systems are inhibited by ammonium ions, which is further evidence that the systems are related. The relationship of ferredoxin in nitrogen fixation is discussed by Yoch and Arnon (1974) and Ljones (1974).

One must conclude this chapter by noting that just as the photosynthetic bacteria have been shown to be more and more diverse in their metabolic pathways, so one might expect the green plant to diversify the range of its products; some indication of this will be discussed in Chapter 10. In biochemistry, however, the bacteria generally show themselves capable of more diverse behaviour and of a greater degree of adaptation to nutritional circumstances than animals and plants. It is nevertheless surprising that the photosynthetic bacteria remain anaerobes, for this limits them to a life in small, isolated habitats, usually on mud under stagnant water, in contrast to the (aerobic) algae which have freedom to live in the open sea.

Suggested further reading

Calvin, M. and J. A. Bassham (1962). *The Photosynthesis of Carbon Compounds*, Benjamin, New York.

Problems

Numerical work, as far as possible based on actual experimental results, covering and extending the material of chapters 1–5. The necessary physical constants, etc., will be found in the Appendix. Answers are p. 198.

Problems

1. A leaf of *Beta vulgaris* may contain 18 μg chlorophyll per square cm of surface area. If it is illuminated by full sunlight (say 0·03 W cm^{-2} visible light energy, of mean wavelength 560 nm) and absorbs 50% of the light, how often does any single molecule of chlorophyll absorb a photon? If the average lifetime of fluorescence of chlorophyll *a in vivo* is 0·7 nsec, what fraction of chlorophyll molecules are excited at any given time?

   Molecular weight of chlorophyll $a = 892$

2. Leaves of *Nicotiana tabacum* L. were subjected to flashes of intense light and very short duration; the fixation of CO_2 was measured as a function of the frequency of flashing (following the principle of the experiments of Emerson and Arnold, see p. 43) and the maximum uptake per flash estimated. The results appeared to be widely scattered but on inspection, and with a sufficient number of repetitions, it was seen that the values were grouped closely round five mean values (see Table, col. I). What could such a distribution mean?

Table

	I			II		
	Normal variety			Mutant strain		
Group	mean value*	S.D.	% of total	mean value	S.D.	% of total
a	4·13	± 0·60	9	3·74	± 0·58	16
b	1·78	± 0·07	20	1·86	± 0·06	46
c	0·99	± 0·03	29	0·94	± 0·05	16
d	0·44	± 0·01	31	0·42	± 0·03	19
e	0·22	± 0·01	11	0·21	± 0·02	3

* mean value of μ moles CO_2 fixed per g chlorophyll per flash. S.D. indicates the standard deviation of each result.

The molecular weight of chlorophyll *a* is 892.

Column II of the table shows the distribution obtained with the *aurea* mutant of the tobacco variety above. What difference is observed? Comparable effects were obtained by varying the age and physiological state of the material used. What conclusions can be drawn from this work?

3. The energy of activation of a chemical reaction can be estimated by measuring the velocity constant at different temperatures, and applying the equation

$$k = Ae^{(-E_a/RT)}$$

where k is the velocity constant at temperature T (absolute), R is the gas constant and A is the Arrhenius constant, specific for the reaction. The delayed light emission (see p. 28) of an algal suspension was measured at temperatures between $10°$ and $40°$, and the relative intensities are given below. If the intensity is proportional to the velocity constant of some chemical or electrochemical process, calculate the energy of its activation. Secondly, given that the emission originates from an energy store in System II, calculate the energy per electron in the store, if the wavelength of the emitted light is centred around 690 nm.

Temp (°C)	$10°$	$20°$	$30°$	$40°$
Relative intensity	1·0	4·1	15.·1	51·3

4. Distinguish between the free energy change (ΔF) for a reaction, and the standard free energy change (ΔF_0).
   When a suspension of chloroplasts was incubated in the light with ADP, ATP and phosphate in a KCl medium containing sufficient ferricyanide and 10 mM Mg^{++} at pH 7·5 and 30°C, the steady-state concentrations of the following reactants after 20 min were: ADP, 10 μM; ATP, 1·7 mM, and phosphate, 2·5 mM.
   (a) What is the free energy change under these steady-state conditions for the synthesis of ATP:

$$ADP + P_i \longrightarrow ATP$$

$$\Delta F_0 = 40·2 \text{ kJ mole}^{-1} \text{ at pH } 7·5, 30°C, 10 \text{ mM } Mg^{++}.$$

   (b) If the synthesis of ATP was coupled to electron flow, what would you expect to be the redox potential span across a coupling site?
   (c) Where, in non-cyclic electron flow from water to ferricyanide ($E_0' = +0·42$ at pH 7·5) is there sufficient energy for the synthesis of ATP at the above concentrations of the reactants?

5. Chromatophores prepared from the photosynthetic bacterium *Rhodopseudomonas capsulata* showed on illumination a red-shift in the absorption spectrum of their carotenoid pigments, measured as a change in extinction at 530 nm. The shift did not take place in the presence of uncoupling agents. What could be inferred from this?
   Chromatophores were suspended in the dark in a potassium-free salt medium (100 mM choline chloride) at 25° in the presence of valinomycin. Valinomycin renders chromatophore membranes permeable to the K^+ ion, while they remain impermeable to all the other ions present. When K^+ was added to the medium (as KCl) a carotenoid shift was seen which was similar to that previously observed on illumination. The table shows the extent of the shift with various concentrations of KCl, and under illumination. What information about the cause of the shift could you derive from this experiment? (Note that the entry of K^+ into the

chromatophores would be opposed by a membrane potential ψ according to the equation

$$\psi = \frac{RT}{F} \ln ([K^+]_{outside} / [K^+]_{inside})$$

where the symbols have their usual meaning, ψ is given in volts and $[K^+]_{inside}$ remains virtually constant.

Table

Conditions	Change in extinction at 530 nm
No additions, steady state of illumination	+0·3
KCl added in dark, to final concentration shown, in the presence of valinomycin: (mM)	
0·1	+0·02
1·0	+0·08
2·0	+0·10
5·0	+0·12
10	+0·15
50	+0·18

6. The radioactivity of $[^{14}C]CO_2$ fixed by the enzyme RuBP-carboxylase appears in the carboxyl-carbon of PGA. By means of a diagram such as Figure 5.1, show the distribution of label among the carbon atoms of the intermediates of the reductive pentose cycle after 1, 2 and 3 turns.

   In an actual experiment the following activities were recorded for the carbon atoms of hexose (a) at 5 sec exposure, and (b) at 30 sec.

	C1	C2	C3	C4	C5	C6
(a)	0·05	0·054	0·73	1·0	0·008	0·008
(b)	0·2	0·17	0·86	1·0	0·14	0·18

   Is this result in accord with your prediction above? Comment.

7. It has been calculated that the chlorophyll content of a single chloroplast is of the order of $2·5 \times 10^{-12}$ gm. Assuming that the chlorophyll content of a spinach leaf is 0·1% of the fresh weight, calculate the number of chloroplasts in a spinach leaf weighing 5 gm (fresh weight).

8. Assuming that you can exist on a diet providing 10^7 J/day and supposing that this requirement could all be met by sucrose (at 17 kJ/gm), how many 5 gm spinach leaves must work 10 hours a day (fixing CO_2 at 100 μmoles/mg chlorophyll/hour) to keep you alive?

10. A sample of 0·1 ml of a suspension of chloroplasts was mixed with 80% acetone at a final volume of 10 ml. The chlorophyll extract was measured in a spectrophotometer with the result

$$E^{1cm}_{645nm} = 0·31$$

$$E^{1cm}_{663nm} = 0·68$$

From the data in the Appendix, determine the concentrations of chlorophylls a and b in the chloroplast suspension.

A further sample of 0·1 ml of the original suspension was added to 2·9 ml of a buffered solution containing 3 mM $K_3Fe(CN)_6$. The mixture was illuminated with a saturating intensity of red light and oxygen production was detected with an oxygen electrode, at a rate of 20 μM per minute. When ADP (0·30 μmoles in 0·1 ml) was added the rate was increased to 39 μM O_2 per minute, for 50 seconds, returning to its original rate. This could be repeated. NH_4Cl (0·1 ml, final concentration 1 mM) was added, and the rate of oxygen production increased to 46 μM O_2 per minute. Finally diuron (DCMU) was added (0·1 ml, final concentration 10 μM) when no more oxygen production was observed at all.

Express all rates of oxygen production and phosphorylation in terms of μmoles (not μM!) per mg of total chlorophyll per hour. Calculate the P/O ratio. Indicate the significance of each observation during the experiment.

Topics for discussion or requiring extended treatment

Discuss the use of the term 'efficiency' in photosynthesis. How can the term be applied (1) to the essentially irreversible reduction of NADP by illuminated chloroplasts and (2) to the conservation of energy in plants?

Comment on the usefulness of attempts to define the 'photosynthetic unit'.

What are the energy requirements for CO_2 fixation? Indicate how these requirements may be provided by green plants.

Describe the measurement of quantum efficiency in the whole cells. Of what value are such measurements?

What is the evidence to support the concept that photosynthesis involves the cooperation of two distinct light reactions? (See Chapter 6.)

Comment on the relationships between the formation of proton and ion gradients and the synthesis of ATP in chloroplasts and chromatophores. (See Chapter 9.)

Compare the overall electron transport pathways in green plants and photosynthetic bacteria. (See Chapter 8.)

What is the evidence that chloroplasts are the site of photosynthesis in plants?

A series of compounds isolated from chloroplasts may be listed in order of decreasing redox potential. Of what value is this list in investigations of energy conservation in plants?

Do the pathways of CO_2 fixation in green plants and photosynthetic bacteria differ in any essential way?

Certain marine molluscs contrive to transfer live chloroplasts from their food plant into their own cells. Would you expect the slugs to obtain a major contribution to their energy from light? What factors need to be taken into account? See Taylor, D. L. (1971). *Comp. Biochem. Physiol.*, **38A**, 233–236.

Part 2

CHAPTER 6

Evidence for two light reactions in photosynthesis in green plants

Emerson and Arnold, in 1932, conceived of the photosynthetic unit as a group of some 2500 chlorophyll molecules coupled to a chemical reaction centre where carbon dioxide was reduced to carbohydrate and oxygen was evolved. Now the position is seen to be more complicated: the reduction of carbon dioxide is seen to depend on an electron transport process, and it is an electron movement itself which is the primary effect of the absorption of light energy. This light-driven movement takes place at two centres possessing some 600 chlorophyll molecules together.

Evidence for these newer ideas accumulated during the decade from 1950, much of it obtained by Emerson himself. The two-light-reaction theory appeared first in 1960 (R. Hill and F. Bendall) and was based on theoretical considerations of the role of the chloroplast cytochromes. In 1961 Duysens and coworkers presented what was to become, essentially, the modern theory, on the evidence of their spectrographic observations; Duysens also allotted the terms photoreaction I, photoreaction II to the NADP-reducing and oxygen-evolving processes respectively.

During the next two or three years the new hypothesis was found to account for and predict many experimental observations. At the same time the 'zig-zag' pathway of electron transport containing the two light reactions in series was found to be the most useful embodiment of the hypothesis. The reader should however bear in mind that while one-light-reaction theories are no longer tenable, leaving the two-reaction principle as the simplest available, it cannot be said with so much certainty that the zig-zag arrangement is the best form of it (see section 8.2). However, no alternative has had a success approaching that of the zig-zag formulation, just as no compelling evidence has been adduced for the existence of three light reactions or more.

In this chapter we shall review the various lines of evidence that today form the basis for our belief that two light reactions are indeed involved in photosynthesis in green plants. We can distinguish four separate topics: first, the observations of differing effects when illumination was varied in wavelength; second, the argument of Hill and Bendall concerning the role of cytochromes and other redox carriers; third, the preparation of sub-lamellar particles by various means from chloroplast thylakoids, which appear to contain separated photoreaction systems; and, lastly, a theoretical argument based on the light energy required to reduce carbon dioxide (or to split water).

6.1 Experiments involving light of varied wavelength

In spite of the undoubted fact that chlorophyll a is the key pigment in both photoreactions of green plant photosynthesis, the proportion of light absorbed by each of them does not remain constant while the wavelength is varied. Had this single principle not been true, we might well have been unable to discover the existence of the two light reactions. The mechanism of the effect may lie to some extent in a different degree of association of the accessory pigments, carotenoids, and phycobilins or the chlorophylls b, c etc., with each of the two types of photosynthetic unit. Probably the more important point however is that chlorophyll a as it occurs *in vivo* is heterogenous, in that various differences of environment or aggregation cause the absorption spectrum to be broadened by the addition of displaced components known as C_{670}, C_{680}, C_{690} and C_{700}. The extent to which these make up the two photosynthetic units may vary from species to species, and during the time course of illumination, as will be discussed in the next chapter. Nevertheless the chlorophyll form that absorbs furthest to the red, C_{700}, can be shown to be almost entirely part of photosystem I, while accessory pigments such as chlorophyll b in higher plants and phycobilins in the blue green and red algae appear to belong more to system II than to system I. It is however true that for the most part the experiments in this section demonstrate that two light reactions are a minimum condition of photosynthesis, but not that two light reactions are sufficient.

6.11 The 'red drop', 'enhancement' and chromatic transients

Early observations showed that the action spectrum (defined in section 2.6) for photosynthesis fell sharply at wavelengths above 680 nm, more steeply than the absorption of the pigments, which continued above 700 nm. Even allowing for the decreased accuracy of absorption and rate measurements at long wavelengths, it was clear than quanta absorbed by chlorophyll above 680 nm were less effective than those of shorter wavelength. This drop in quantum efficiency was termed the 'red drop', and an example is given in Figures 2.7(b) and 6.1.

In Figure 2.7(b), it was shown that the quantum yield of photosynthesis in whole cells of *Navicula* dropped sharply at wavelengths above 670 nm. In Figure 6.1 isolated chloroplasts were used (note that the ordinate of the graph is the *quantum requirement*, the reciprocal of the quantum yield) and two conditions were applied. Using $NADP^+$ as the electron acceptor, and water as the donor, the quantum requirement (quanta per electron transferred) rose slowly around 660 nm and rapidly above 680 nm. However, when an artificial donor system was used, the curve parted company with the first at approximately 670 nm and fell to a value close to unity, the theoretical minimum. The explanation advanced for this effect is that electrons from water have to pass through two reaction centres to reach $NADP^+$, and therefore the lowest possible quantum requirement is 2 (that is, the maximum quantum yield is $\frac{1}{2}$). Moreover, the reaction can only proceed as long as both reaction systems are working; at long wavelengths system II absorbs a diminishing proportion of the total light, so that the rate declines, and the greater part of the absorbed energy (in system I) is wasted.

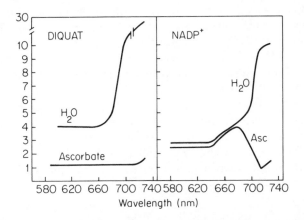

Figure 6.1. The quantum requirements for photosynthetic electron transport from water *or* ascorbate to NADP *or* diquat (a viologen compound). Note the sharp rise in the requirement when water is the donor, also the apparent change in the relative distribution of energy under the two sets of conditions. From Avron and Ben-Hayyim (1969), with permission

Hence the observed red drop in the quantum yield. Using ascorbate as a donor, the experimenters found that in the long wavelength region where most light was absorbed by system I, there was less light wasted and hence the artificial donor bypassed system II. At wavelengths where system II did absorb, that part of the light was wasted, so the quantum yield requirements were the same as with water. In this case there did not appear to be any provision for the chloroplast to transfer energy from one photosystem to the other.

However, when an artificial acceptor (the viologen herbicide diquat) was used instead of NADP, the quantum requirement for electrons from water was increased (the loss of efficiency above 670 nm being still present) while that for electrons from the artificial donor was constant at a value just above unity. This was used by the authors to argue that the distribution of chlorophyll between the two photosystems could be affected by the experimental conditions, and hence possibly by conditions inside the cell, allowing some degree of control. This is a concept of 'plastic' chlorophyll that can change its organization in response to circumstances, and it will be discussed later.

Emerson showed with algae that the quantum yield of light in the far-red region was enhanced by a weak background illumination with shorter wavelength light. Enhancement was said to occur when the rate with the two beams of light together was greater than the sum of the rates with either beam alone. The effect, also known as the Second Emerson Effect (the first was concerned with carbon dioxide 'gush' occurring with intact algae at the onset of illumination) has been well documented for algal cells, and for chloroplasts with intact envelopes, when either oxygen evolution or carbon dioxide fixation was measured. Some interesting results have been obtained by the study of *enhancement spectra*: these display the degree of enhancement of a fixed wavelength background illumination by a variable-wavelength beam. The resulting spectra show peaks which can be reasonably well identified with the pigments of the cells. Figure 6.2 (Fork, 1963) shows enhancement spectra for several species. It can be seen that chlorophylls *b* and *c* enhance the absorption by chlorophyll *a*, and also that the blue-absorbing peak of chlorophyll *a*, at 436 nm follows the far-red absorbing band at 690–700 nm.

Enhancement is usually measured by selecting two wavelengths, λ_1 and λ_2, and obtaining the rates of reaction with each beam singly ($R\lambda_1$, $R\lambda_2$), and with both together ($R_{(\lambda_1+\lambda_2)}$). The value for enhancement may be given either as a percentage (formula 6.1), where the condition of no-enhancement is a zero result, or as a simple ratio (formula 6.2), where the same condition is indicated by the value 1·0:

$$\text{Enhancement} = \frac{(R_{(\lambda_1+\lambda_2)} - R_{\lambda_1} - R_{\lambda_2}) \times 100}{R_{\lambda_1} + R_{\lambda_2}} \tag{6.1}$$

$$\text{Enhancement} = \frac{R_{(\lambda_1+\lambda_2)}}{R_{\lambda_1} + R_{\lambda_2}} \tag{6.2}$$

The values obtained are small, being under most conditions less than 2 or 3 (formula 6.2). This is an indication that quanta of wavelength less than say 670 nm are efficiently distributed between the two photoreactions by a variable spillover mechanism.

The third group of experiments in this section is the measurement of *chromatic transients*; these are sudden and shortlived jumps or drops in the rate of, say, oxygen evolution when the wavelength (but not the effective intensity) of the illumination is suddenly changed. This is the Blinks effect (Blinks, 1957) and was pursued intensively by French and Myers. This effect can be considered as an Emerson-enhancement experiment in which the two wavelengths are applied successively rather than simultaneously. The dark interval between the two light beams may be extended, when the magnitude of the transient is found to diminish, suggesting that a chemical intermediate accumulates by the operation of one light-reaction, and is removed by the other. Quantitatively the experiment is hard to evaluate, since many effects seem to be involved. Thus in whole algal cells there is a light-stimulated uptake of oxygen (photorespiration) and under some conditions the production of oxygen is preceded by a temporary uptake.

6.12 Antagonistic effects of light beams of different wavelengths

The observations of the previous section showed that two light beams of

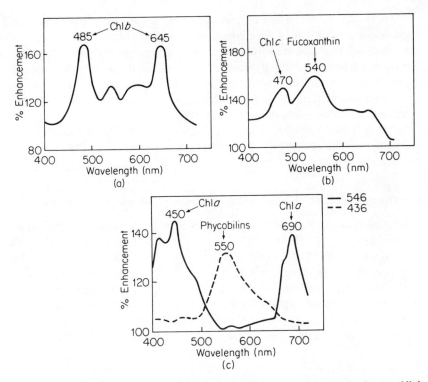

Figure 6.2. Enhancement in green, brown and red algae. (a) *Ulva*, with background light of wavelength longer than 680 nm. Note the prominence of the bands of chlorophyll *b*. (b) *Endarachne binghamiae*, with a background as in (a). Note the contribution of Chl *c* and fucoxanthin. (c) *Porphyra perforata* with two conditions of background. From Fork, D. C. (1963) In *Photosynthetic Mechanisms of Green Plants* Publ. No. 1145 NAS-NRC Washington, D.C. p. 352, with permission

differing wavelengths cooperated in the process of photosynthesis. The present section is concerned with an apparent antagonism. Thus illumination by far-red light (700–720 nm) causes the oxidation of cytochrome f*, plastocyanin, P700 and plastoquinone; these materials become more reduced under shorter-wavelength light. In addition, far-red light acts to diminish the degree of fluorescence obtained at room temperature from flashes of light at 620–680 nm. The level of fluorescence emission from system II pigments is thought to be under the control of a 'quencher', Q, which is inactivated and activated reversibly; in the absence of far-red light, light of 620–680 nm produces immediate fluorescence at a level indicative of the state of Q at the time, and the emission increases in intensity up to a maximum as Q is inactivated. (Further, slower changes occur after that, which Duysens considers to indicate a rearrangement of chlorophyll molecules from system II units to system I.)

---

* The observations on cytochrome f were the basis of the distinction between systems I and II originally made by Duysens and coworkers (1961).

104

These results are compelling evidence that there are at least two separate kinds of photoreaction centres in the chloroplasts of higher plants, and that there are significant differences in the absorption spectra of the pigments attached to them. The zig-zag scheme provides a convenient rationale of the observed antagonistic effects of the two wavelengths: if the two light reactions and the redox carriers are in a linear series, and if one of the light reactions becomes rate limiting, the carriers on the source-side of it tend to become more reduced, and those on the acceptor-side more oxidized. Changing the wavelength of the illumination may cause the other light reaction to become rate-limiting instead, and the state of the carriers will change accordingly.

6.2 The need to accommodate redox materials of intermediate potential

The chloroplast cytochromes, f, b_6 (both 559 and 563 nm components) and the high potential cytochrome-559 all have standard redox potentials at pH 7 in the range 0–0·4 V. Plastocyanin, P700 and the plastoquinones also have standard potentials in or close to that bracket. Probably the first clear suggestion of the zig-zag scheme was that given by Hill and Bendall (1960b), on the basis of the argument that any one-light-reaction theory accounting for electron transport from water (+0·82 V at pH 7) to NADP (−0·32 V) would only use electron carriers with potentials outside that range, as indicated in Figure 6.3. (Remember

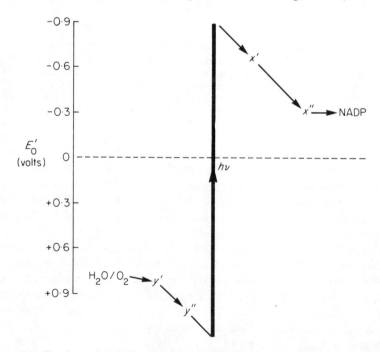

Figure 6.3. To show that redox materials involved in a one-light-reaction electron transport process from water to NADP must have E_0' values outside the range +0·82 V–0·32 V

that in these diagrams spontaneous electron flow is represented by a downward arrow, upward arrows indicating a light-driven step.) Hill and Bendall showed that cytochromes b_6 (0·0 V) and f (0·37 V) could be accommodated by a two-light-reaction scheme (Figure 6.4) There were at that time some indications that f

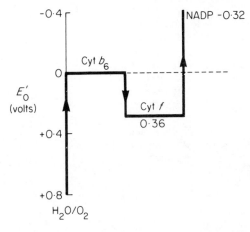

Figure 6.4. The first formulation of the 'zig-zag scheme'. From Hill, R. and F. Bendall (1960), *Nature*, **186,** 136 with permission

was oxidized and reduced by illumination under various conditions, and these have since been well established. A further reason for involving cytochromes was that the phenomenon of photophosphorylation had been recorded by Arnon's group, and Hill and Bendall were conscious of the possible analogy between that and oxidative phosphorylation. An intimate account of the development of these concepts has been given by Hill (1965), and also by Arnon (1968).

6.21 The existence of two series of electron acceptors

Kok (1966) has given an account of various electron acceptors in photosynthetic electron transport (Hill oxidants). Some of these can be arranged into two series. One series, the viologen dyes, have low potentials ($-0·318$ V to $-0·740$ V) and are reduced by chloroplasts according to Table 6.1, where it can be seen that there is some limiting potential at approximately $-0·65$ V. Zweig and Avron (1965) have also reported a similar limit, at $-0·5$ V, and Black (1966) at $-0·52$ V. A second series, of substituted quinones, was applied to a mutant alga, *Scenedesmus* No. 8, which was unable to carry out the far-red photoreactions of photosynthesis. The quinones were reduced by the Hill reaction, down to a limit, this time at a potential of $+0·17$ V. A more sensitive method was found, in which Kok and his colleagues measured the intensity of fluorescence (which comes mainly from system II at room temperature, and which is governed by the redox-state of the quencher, Q). By adjusting the concentrations of oxidized and reduced naphthoquinones, making redox buffers at various potentials in which

Table 6.1. Two series of Hill reagents

Series I—viologens			Series II—quinones	
E_0(mV)	% reduced		E_0(mV)pH 6·5	O_2 evolution
−318	100	$K_3Fe(CN)_6$	+420*	+
−440	100	p-Benzoquinone	+330*	+
−548	100	Methyl-p-benzoquinone	+280	+
−670	26	2,5-Dimethyl-p-benzoquinone	+220	+
−740	4	2,6-Dichlorophenol-indophenol	+220	+
		1,2-Naphthoquinone	+190*	+
		1,2-Naphthoquinone sulphate	+160*	weak
		1,4-Naphthoquinone	+90	−
		2-Methyl-1,4-naphthoquinone (menadione)	+20	−
		2,5-Dihydroxybenzoquinone	−80	−
		2,Hydroxy-1,4-naphthoquinone	−90	−
		Sodium anthraquinone sulphonate	−180	−
		FMN	−180	−
		Methyl viologen	−440	−

* At pH 6·6.
Data from Kok (1966) and Kok and Datko (1965).
The dotted lines, drawn through the list of reagents, indicates the point in each series where the potential is too low (negative) for electrons to be accepted from system I and II respectively.

chloroplasts were suspended in the dark, the initial level of fluorescence was found to vary, indicating that Q responded to the redox potential of the solution. Since there should have been equilibrium in the dark period, the external potential at the point of 50% quenching of fluorescence indicated the redox potential of Q: a value of +0·18 V was obtained. Cramer and Butler (1969) have extended this approach and have produced values for the redox potential of Q of between −20 mV and −35 mV (see section 8.22). It was necessary to use the mutant since otherwise the low-potential reductants produced by photoreaction I would have interfered with the higher potential product of reaction II. It can be concluded that the electron acceptors were reacting at two distinct sites, which had widely differing redox potentials. The more negative of these is low enough to account for the reduction of NADP, but the more positive potential has no explanation on a one-light-reaction theory. If on the other hand the two-light-reaction hypothesis is accepted, these measured potentials give an indication of the potentials of the primary reductants produced by the two photoreactions: −0·5 V (system I) and −0·03 V (system II).

6.3 Fractionation of the thylakoid membrane

Treatment of the chloroplast with surface active agents often results in its disruption into small sub-lamellar particles. Thus Boardman and Anderson found that *digitonin* allowed the separation by differential centrifugation of two fractions, one containing small particles, with a higher percentage of chlorophyll *a*, which were more active in the reduction of NADP by ascorbate, and a residue of much larger particles, having a lower ratio of chlorophyll *a/b* than in the original chloroplasts and which was the more active in the Hill reaction (the production of oxygen with ferricyanide as electron acceptor). Since the total recovery of the two activities was good, it was claimed that the digitonin had physically divided the thylakoid into constituent parts, and that each part clearly carried out a different photoreaction. This work has been developed by several workers and will be reviewed in the next chapter. For our purposes here, it can be said that particles which carry out the ascorbate-NADP photoreaction (I) have been prepared by several methods and refined to small size, containing little more than protein, chlorophyll *a* and P700. However for the full proof the same should be done with the Hill reaction, and this appears to be impossible, because once the thylakoid is broken down to small particles, apparently by any means, the Hill reaction is lost. There is some reason to believe that the Hill reaction requires a vesicular structure and the smallest particles prepared that have been Hill-active were vesicles of some 50 nm diameter. Smaller particles than that are easy to obtain, but are inactive.

6.4 Energetics and the quantum yield

From the second law of thermodynamics, any apparatus which irreversibly converts one form of energy into another must show a degree of inefficiency, that is, the energy stored must be less than the energy supplied. The difference appears as liberated heat (as in the effect of friction) and as an increase in disorder. A photosynthetic theory must allow for this law: the total energy of the quanta of light absorbed must be more than the total energy change represented by the chemical reaction carried out.

Light of 680 nm has an energy of approximately 172 kJ (41 kcal) per einstein. Since the standard free energy change in the reduction of carbon dioxide to carbohydrate, with the production of oxygen from water is 502 kJ (120 kcal mole^{-1}), 3 quanta per molecule of carbon dioxide would achieve a balance, and 4 quanta would give a figure for the efficiency of the system of $502/688 = 73\%$, which is high (but not inconceivable). Since however we know that the complete chemical process involves many steps, we can refine our estimate by considering the energy trapped by the primary photoprocess itself. For this purpose the energy of the 680 nm quantum is better expressed as approximately 1·8 electron-volt (eV) per photon. The difference in the redox potentials of the primary electron donor(s) and acceptor(s) should not exceed 1·8 V for a one-quantum process. There is a reasonable estimate of the potential achieved at the reducing side, based on observations of the degree to which chloroplasts will reduce artificial (viologen) electron acceptors: the value is approximately −0·52 V. The oxidizing potential achieved is difficult to estimate since the primary donor is now known; but it must be at least (say) 100 mV more positive than the standard potential of the oxygen–water couple ($+0.82$ at pH 7) since there seems to be no tendency for the evolution of oxygen to reverse. The minimum difference between the two potentials is 1.44 V to which we must add the energy stored as ATP. The standard free energy of ATP formation from ADP and P_i under the conditions of pH, temperature and magnesium-ion concentration reasonable in a chloroplast has been estimated at 30 000 J mole^{-1} (7.4 kcal mole^{-1}) which is equivalent to 0.3 eV per molecule. However, photophosphorylation does not normally run at equal concentrations of the reactants and products; it appears that ATP can be synthesized until the ADP concentration is very small indeed. The actual free energy of formation at this point may be perhaps double the standard figure. It is unfortunately the case that there is no agreement on the *possible* number of ATP molecules synthesized during the passage of one pair of electrons through systems II and I, as will be shown in Chapter 9. For the present, taking an upper but not impossible estimate of 2, we allow 0.6 eV for ATP formation on top of the potential difference between the termini of the electron transport pathway (1.4 V, depending on the estimate of the potentials of the electron-acceptor of the watersplitting reaction and the donor in the reduction of ferredoxin). The total considerably exceeds the energy available from one quantum of red light (1.8 eV), and therefore more than one quantum is required.

Even if we suppose that phosphorylation is less complete in the system involving artificial electron acceptors (there is little reason why this should be so) or take the lower value for the ATP/2e$^-$ ratio, we still obtain a total energy storage approaching the 1.8 eV supplied, sufficiently close to throw doubt on a one-quantum mechanism.

Another way of looking at the matter is to consider the quantum yield measurements for photosynthesis, in algal cells. Many workers have consistently measured maximum yields of about 0.1, sometimes as high as 0.11, molecules of oxygen evolved per photon absorbed (see Emerson, 1958). It is more satisfying to consider 0.11 as an experimental approximation to 0.125 indicating that eight quanta are required for the four-electron process, rather than to suppose that the efficiency of the pigment system itself is less than 50%.*

If it is accepted on the evidence reviewed in this chapter that there are two light reactions in photosynthesis, then the observations that follow can be applied to the discussion of the form the mechanism must take. We shall examine the structure of the thylakoid more closely, and then review electron transport and the evidence concerning the choice of the zig-zag or other models.

---

* Warburg for many years claimed that yields of some 0.25 or higher can be obtained under the right conditions. If this is the case, the two light reaction theory is weakened severely; however most workers consider these high values to be experimentally unsound. A paper by Emerson and Nishimura (1949) indicates the ground on which this criticism is usually based. Warburg has on the strength of his measurements put forward a one light reaction theory (section 8.24).

CHAPTER 7

The structure of the thylakoid membrane

The thylakoids of the chloroplast present a most striking appearance in electron micrographs, of closely appressed pairs of membranes, often oriented together and stacked; the membranes are thicker, and stain more densely with most electron-dense stains than membranes from other organelles and tissues. Membranes currently form a very active topic of biochemical research, for three general reasons: they constitute a stable, solid phase, two-dimensional system in the liquid-phase, three-dimensional medium of the cell, and the materials of which they are made have interesting physical and chemical properties; secondly, since a membrane forms a barrier which resists the passage of many solutes and supports electric potentials, we may look for some special organization of the protein and lipid components; thirdly, virtually all subcellular membranes have a degree of catalytic activity, from the facilitated-diffusion systems involving 'permease' enzymes, and the existence of pores for the entry of specific materials, to active transport systems that perform osmotic work at the expense of ATP, and electron-transport systems that can conserve the energy of redox reactions.

The structure of the thylakoid membrane attracts interest not only in connection with the above, but also by virtue of the heterogeneity implied by the two light reactions of photosynthesis. We have seen that there are units of some 600 chlorophyll molecules partly or completely divided between the two reaction centres, and we might expect to find particles, one or perhaps two per 600 chlorophylls. Again, we ascribed differences in the reaction spectra of the two photoreactions to differences in the pigment types associated with each centre; if the pigments are functionally distinct we might expect to find a structural separation. Such a particulate understructure might be found either by fragmenting the thylakoid, or by using the electron microscope; both methods have yielded interesting results.

Membranes in cells, such as the cell envelope, the endoplasmic reticulum and the outer and inner membranes of mitochondria and nuclei may have an underlying similarity. Thus possession of cytochromes has been demonstrated not only in mitochondrial inner membranes, but also in nuclear membranes and the endoplasmic reticulum, so that one might wonder if traces of electron transport carriers might be found in the cell membrane and in the chloroplast outer envelope. Although the inner membranes that form the cristae of the mitochondrion and the thylakoid of the chloroplast are thicker than the others, there is evidence that these membranes are formed by proliferation of the outer membrane of the organelle, the thickening occurring subsequently. This would imply a fundamental homology in the structures of thick and thin membranes. There are also observations that suggest that mitochondria and chloroplasts develop as outgrowths of the nuclear membranes, although discussion of this point is beyond the scope of this text.

7.1 Structural components

It will be obvious at the outset that the label 'structural component' tends to be applied to those materials for which no other function has been found. At the same time the structural properties of catalytically-active components are often overlooked. Table 7.1 (from Lichtenthaler and Park, 1963) gives a list of the principal components of thylakoid fragments, and their abundance

Table 7.1. Representative distribution of substances in spinach chloroplast lamellae on basis of minimum molecular weight of 960 000 per mole of manganese (From Lichtenthaler and Park (1963), with permission).

Lipid (composition, moles/mole Mn)		
115 chlorophylls		103 200
80 chl. *a*	71 500	
35 chl. *b*	31 700	
24 carotenoids		13 700
7 β-carotene	3 800	
11 lutein	6 300	
3 violaxanthin	1 800	
3 neoxanthin	1 800	
23 quinone compounds		15 900
8 plastoquinone *A*	6 000	
4 plastoquinone *B*	4 500	
2 plastoquinone *C*	1 500	
4,5 α-tocopherol	1 900	
2 α-tocopherylquinone	1 000	
2 vitamin K_1	1 000	
58 phospholipids		45 400
(phosphatidylglycerols)		
72 digalactosyldiglyceride		67 000
173 monogalactosyldiglyceride		134 000
24 sulpholipid		20 500
? sterols		7 500
Unidentified lipids		87 800
		————
		495 000
Protein		
4690 nitrogen atoms as protein		464 000
1 manganese		55
6 iron		336
3 copper		159
Lipid + protein		————
		465 000*
		960 000

* Rounded off to nearest thousand.

(plastocyanin may well have been partly lost in the preparative procedure), from which it can be seen that the pigments occur in sufficient quantity to make a considerable contribution to the properties of the water-insoluble part of the matrix. Proteins such as the apoprotein of cytochrome f (molecular weight 62 000 per haem group) could be playing an additional, structural, role. However there is little doubt that a major structural feature is the 'structural protein' which accounts for some 20–30% of the dry weight of the thylakoid. Criddle (see for example Criddle, 1966) prepared protein of this kind first from mitochondria, and then from chloroplasts, demonstrating the similarity. The method Criddle used was to remove the pigment from the thylakoids with acetone, and extract the dried powder with cholic and deoxycholic acids, finally fractionating the product with ammonium sulphate and sodium dodecyl sulphate (SDS). 'Structural protein' has also been obtained from the erythrocyte 'ghost', although this last differs from the others in that it is soluble outside much narrower pH limits. At moderate pH values the chloroplast protein is only soluble in detergents such as cholic acid, or sodium dodecyl sulphate. Structural protein of both chloroplast and mitochondria is characterized by a molecular weight in the region of 23 000, and a high proportion of non-polar amino acids. Criddle showed that it was able to bind many porphyrin-type compounds in stoichiometric proportions: chlorophyll, myoglobin and cytochromes. The detergent–chlorophyll–protein particles prepared by Thornber and coworkers (1967b) can be regarded as two structural proteins with different amino acid compositions. The Criddle preparation may be a mixture of these two proteins.

The thylakoid has its complement of lipids, which are characteristic. Table 7.1 contains a list of these. The unsaturated fatty acids are interesting both for their abundance (said to be a feature of oxygen-evolving photosynthetic systems), and from their resistance to oxidation by the high partial pressure of oxygen which must obtain during photosynthesis. The resistance is often lost when chloroplasts are isolated. Apart from noting the existence of the characteristic thylakoid lipids (of which some formulae are given in Figure 7.1), and commenting that they may have an insulating and cementing-role in the membrane, there is little that can be said concerning their arrangement. Weier and Benson (1966) have however prepared for discussion a model structure in which various types of lipid are allocated a position in the thylakoid.

Thylakoid–thylakoid binding. It should be noticed that in grana the outer surfaces of thylakoids adhere together, and continue to do so if the granum is osmotically swollen; the swelling takes place inside the thylakoid. The outside of the thylakoid (in the granum) was thought to be hydrophobic, in contrast to a hydrophilic interior. It now seems clear, however, that the stacking of thylakoids is related to the binding of cations, since Izawa and Good (1966) found that the degree of stacking could be regulated by the cationic composition of the medium. This implies that cations can penetrate between thylakoids, in the *partition-gap*, which therefore contains water. Electron microscopy has confirmed that the space between thylakoids is an extension of the stroma (see the review by Anderson, (1975), p. 222). The mechanism of photosynthesis would be even harder to understand if water were excluded from the partition gap!

Stacking has been correlated with the presence of the chlorophyll–protein complex II ($a + b$), and lamellae from the stroma, or from agranate chloroplasts, which have less of this complex, do not aggregate even in high salt solutions. Preparations of the complex, free from detergent, form aggregates, and the speed

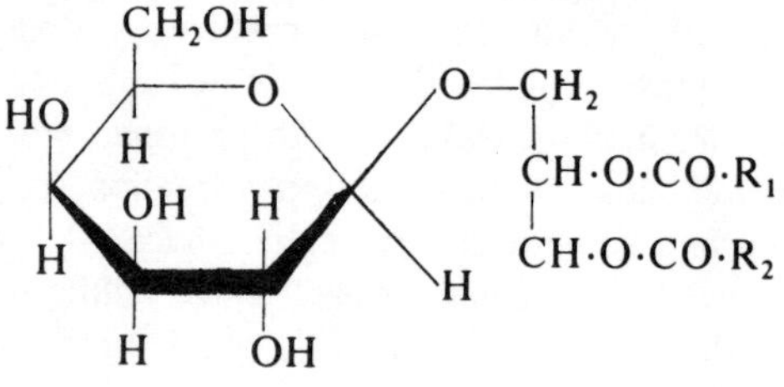

I Monogalactosyl lipid

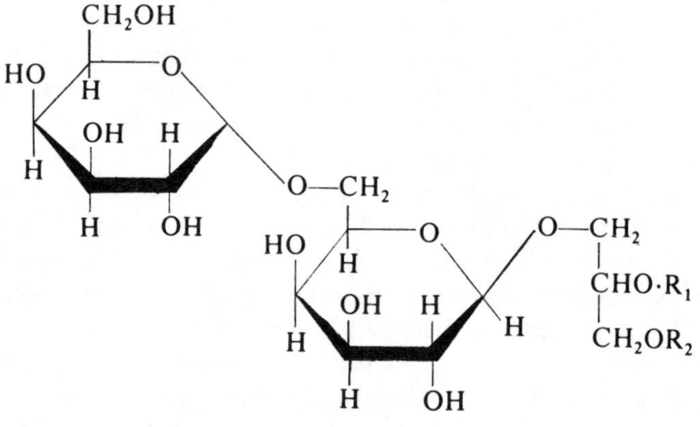

II Digalactosyl lipid

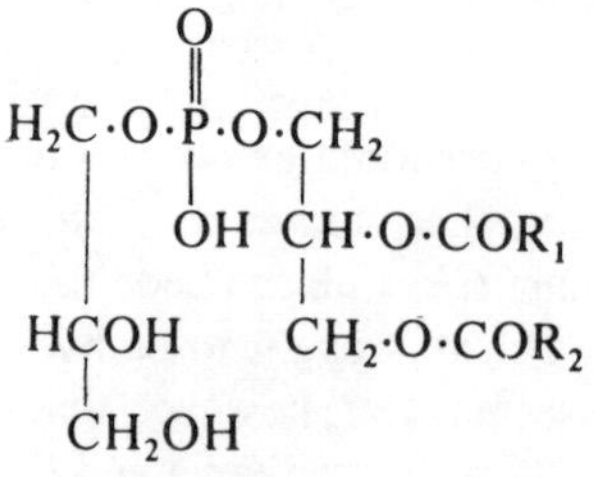

III Phosphatidyl glycerol

Figure 7.1. Formulae of some of the thylakoid lipids, from Weier, T. E. and A. A. Benson (1966). In T. W. Goodwin (Ed.) *Biochemistry of Chloroplasts*, Vol. 1, Academic Press, London, p. 91, with permission

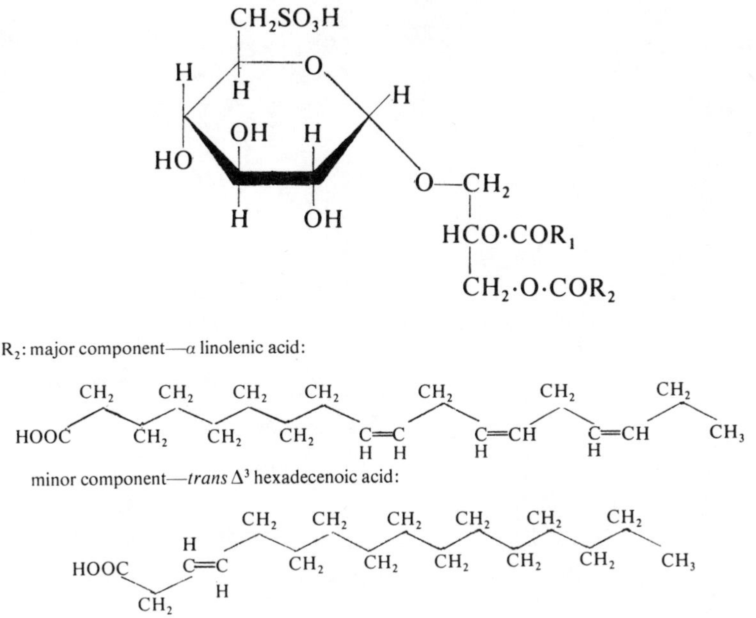

CH₂SO₃H structure:

$$CH_2SO_3H$$

(figure showing sulphoquinovosyl diglyceride structure with labels H, O, OH, H, HO, H, O—CH₂, HCO·COR₁, CH₂·O·COR₂)

R₁, R₂: major component—α linolenic acid:

(structure diagram with labels)

CH₂   CH₂   CH₂   CH₂   CH₂   CH₂   CH₂

HOOC   CH₂   CH₂   CH₂   C=C   C=CH   C=CH   CH₃
                                    H H   H   H

minor component—*trans* Δ³ hexadecenoic acid:

                    CH₂   CH₂   CH₂   CH₂   CH₂   CH₂
            H
HOOC   C=C   CH₂   CH₂   CH₂   CH₂   CH₂   CH₃

    CH₂   H

IV Sulphoquinovosyl diglyceride

Figure 7.1. continued

and extent of aggregation is greater in high salt concentrations (Scott, 1974; N. al Sakkal, unpublished). The complex might therefore provide specific bridges, linking thylakoids in the granum. Electron micrographs (in the author's estimation) leave the question of the existence of bridges open. It is hard to see how else the membranes could be held so precisely together, and also kept apart.

7.2 Pigments

There is a degree of organization shown in the complement of pigments of photosynthetic structures. We are led to this conclusion by four lines of evidence: (i) the conclusions of Chapter 6, that there were two photoreactions, drawing their energy from distinguishable groups of pigment molecules; (ii) resolution of sub-lamellar particles which differ in the proportions of the chlorophylls *a* and *b* and carotenoids (see section 7.4); (iii) optical measurements of the phenomena of dichroism, birefringence, linear dichroism, optical rotatory dispersion, circular dichroism and polarized fluorescence that can detect various modes of organization; and (iv) the resolution of chlorophyll (in the intact chloroplast) into discrete components that are identical chemically but have different absorption properties.

Optical effects. Consider an isolated chlorophyll *a* molecule. Absorption of light of 680 nm leads to an electron-displacement within the plane of the tetra-pyrrole

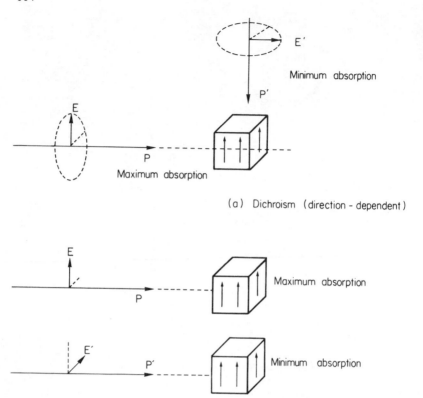

(a) Dichroism (direction - dependent)

(b) Linear dichroism (polarization - dependent)

Figure 7.2. Diagram to illustrate the terms dichroism and linear dichroism. In (a) rays propagated in the direction P, P' and having electric vectors in planes E, E' are incident on a crystal: absorption cannot take place if the electric vector is at right-angles to the absorption dipoles indicated by the arrows in the crystal. Hence the absorption is direction-dependent. (b) Two rays polarized vertically and horizontally (E, E' as shown) will be absorbed differently by the crystal if the orientation of the absorption dipoles in parallel with E and at right-angles to E'

ring, along the axis of rings I and III. This direction is the direction of the absorption dipole. Since the absorption takes energy from the electric field of the light wave, the electric vector of the wave must be parallel with the absorption dipole for the maximum probability of absorption. Consider next an array of chlorophyll molecules. To a first approximation the incident light ray interacts with each molecule separately. The total absorption will depend on the number of molecules encountered, and their average orientation with the electric vector of the ray. If there is a significant orientation of the absorption dipoles in a given direction, then it is to be expected that the absorbance will depend on whether the direction of the propagation of the ray is parallel with that direction or transverse to it. (The electric vector is at right-angles to the direction of propagation of the ray, see Figure 7.2.) The differential absorption of such an array of molecules is called *dichroism*. It is more usual to control the direction of the electric vector not by the direction of the ray but by means of *plane polarization*. Absorption will be greatest when the mean absorption dipole lies in the plane of polarization. Differential absorption of two beams, polarized at right-angles to each other, is called *linear dichroism*, although the two terms are not always distinguished.

Absorption phenomena at a given wavelength are related to the behaviour of a few electrons. All the electrons of a molecule, however, contribute to the refractive index, in proportions that depend on wavelength. The *polarizability* of a molecule at a given wavelength may be vectorial in the same way as the absorption dipole. An array of oriented molecules may have two refractive indices depending on the direction of the incident ray or the plane of its polarization (see Figure 7.2). The existence of two different refractive indices is *birefringence*. An unpolarized ray incident in a certain direction on a birefringent crystal will be split into two emergent rays, plane-polarized at right-angles to each other. This is *double refraction*.

With chloroplast grana, the parallel membranes can themselves generate birefringence, called *form* birefringence (as opposed to the *intrinsic* birefrigence of oriented pigments). This leads to differences in the light scattered from the membrane–stroma interfaces. In many spectrophotometers scattered light is lost and is treated as if it had been absorbed; hence form birefrigence can lead to a false linear dichroism. This can be checked by varying the refractive index of the medium (using sucrose for example). When the refractive index of the medium is equal to that of the membranes, form birefringence vanishes. This is described by Thomas and coworkers (1966) in measurements of linear dichroism in spinach chloroplasts. The technique is further complicated by the severe changes in the refractive index of thylakoids in the region of the absorption bands of chlorophyll (anomalous dispersion) which can lead to selective scattering.

A plane-polarized beam may be resolved, both theoretically and experimentally, into two components which are circularly polarized, right- and left-handedly. Circular polarization may be visualized by considering the electric and magnetic vectors, and the propagation direction, as three mutually orthogonal directed axes, using the thumb, forefinger and second finger of the hand for illustration; the two hands show the two forms of circularly polarized light. It should be remembered however that the vectors are not confined to planes as in plane polarized light; it is the symmetry that is defined. Chromophores which are asymmetrical, either intrinsically, or owing to the asymmetry of the molecule as a whole, have different absorption coefficients for each kind of circularly polarized light. This is *circular dichroism* (CD). (The difference is typically of the order of magnitude of 1 in 10^4.) The corresponding effect of the difference in refractive indices is responsible for the rotation of the plane of plane-polarized light by 'optically active' materials, which when measured as a function of wavelength is known as optical rotatory dispersion (ORD). Both CD and ORD are affected if the pigment molecule is influenced by an asymmetric field, such as a protein molecule could generate, or by aggregation of the pigments into arrays with their own degree of asymmetry.

Chloroplasts show birefringence and dichroism: Rabinowitch (1945) cites Scarth as the original discoverer of the birefringence in 1924, the phenomenon being rediscovered by several workers ten to twelve years later. Of these, Menke reported dichroism in 1938, giving the wavelength of maximum effect as 681 nm. Olson (1963) observed dichroism in intact chloroplasts using an infrared image converter and a microscope. With monochromatic light he observed dichroism with a maximum in the range 695–705 nm, but only at the edge of the chloroplasts (that is, with the thylakoids in profile). There was linear dichroism also, the electric vector being tangential to the thylakoids as seen at the edge of the chloroplast for maximum absorption. This suggested that only a part of the total chlorophyll was organized in any perceptible oriented structure.

This view received support from observations of *polarized fluorescence*. Most fluorescence from chloroplasts is at 685 nm (at room temperature) but there is a small emission around 720 nm. When chloroplasts were illuminated with a mercury lamp (436 nm, convenient since the transmitted light can be stopped by means of a red filter) red fluorescence was observed, unpolarized except for the 720 nm fraction. The plane of the polarized fluorescence at 720 was tangential to the edges of the chloroplasts as above. A variation of the above experiment was carried out in which light from a ruby laser (694·3 nm) which is plane-polarized was used to excite the 720 nm fluorescence. The ruby light is of too long a wavelength to excite much 685 fluorescence, and the 720 nm emission was found to vary in intensity according to the plane of polarization of the laser. The conclusion was generally accepted that the chlorophyll is randomly arranged, except for the small (5% or less) fraction absorbing in the range 695–705 nm. However, this does not follow; the results above indicate only that there is a sufficient proportion of random chlorophyll in the main absorption band to allow for the direction of polarization to be lost in the number of jumps an excitation makes between the absorbing and emitting molecule. Direct measurements of the degree of orientation of the absorbing molecules were made by observing the linear dichroism (defined above) of chloroplasts that had been aligned, either by spreading on a transparent plate (as at Saclay, by Breton and coworkers, 1973) or by the action of a magnetic field (as at New York University by Geacintov and coworkers, 1972). These results showed that a large proportion of chlorophyll *a* absorbing at 681 nm (some 60%) was oriented so that the transition moments were inclined at an angle to the plane of the thylakoid membranes. This fraction of chlorophyll is also implicated in the intense circular dichroism effect studied by Gregory and Raps (1974). This signal which had some of the properties expected of a close interaction between chlorophyll molecules, was found to be associated with the apposition of thylakoids in grana (Faludi-Daniel and coworkers, 1973). Gregory (1975) found a dependence of the circular dichroism on the orientation of the thylakoids brought about by passing the suspension of the chloroplasts through a flow cell.

All the above lines of research suggest that there is a considerable degree of interaction and orientation in chlorophyll *in situ*.

We have had occasion to distinguish part of the thylakoid chlorophyll by means of its absorption and fluorescence maxima. The idea that chlorophyll in the chloroplast was heterogeneous appears to have begun with Albers and Knorr, who in 1937 showed that inspection of the absorption spectra of certain preparations revealed the presence of components with maxima at 674 and 683 nm, and possibly others at 668, 687 and 698 nm. Since the overall absorption peak, in the red, of chlorophyll *in vivo* is so much broader than in solution, even after allowing for the effects of light scattering, this was an attractive concept, and the work has since been confirmed many times. French and his coworkers have recently set up a computer routine for fitting idealized

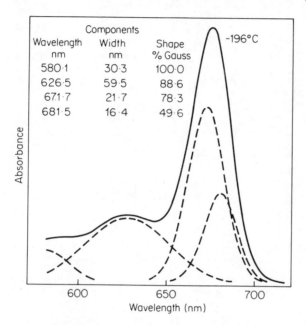

Figure 7.3. The absorption spectrum of a pale mutant of *Chlorella*, from which a computer program has extracted the 'components' as shown. From French, C. S. and L. Prager (1969). In H. Metzner (Ed.) *Progress in Photosynthesis Research,* Institut für Chemische Pflanzenphysiologie, Tubingen, p. 558, Fig. 2, with permission

chlorophyll spectra to the actual spectra of algal cells. An example of the kind of result obtained is given in Figure 7.3. Previously the same group made considerable progress with a differentiating spectrophotometer; this instrument recorded, instead of the extinction, E, the derivative $dE/d\lambda$, against the wavelength. This has proved easier to interpret than the absorption curves themselves, at least without computer assistance. An example is given in Figure 7.4. Work of this kind has been carried out in several laboratories, and a consensus has grown up that there are at least three components, the first two being major, each constituting close to 50% of the total chlorophyll. The wavelength ranges of these two are generally agreed to be 670–673 nm and 680–683 nm; both the precise wavelength and quantity differ from species to species, and in different preparations. The third type absorbs variously in the range 695–705 nm or even beyond. Opinion is at present divided on whether this is a single class or not; we will argue here that it makes better sense if the 695 and 705 nm components are separate. The 695–705 nm form(s) constitute some 3–5% of the total chlorophyll. We should distinguish also the photoreactive pigment P700, which is reversibly bleached by light or by oxidation, and is believed to be chlorophyll at the active centre of system I. Kok, the discoverer of P700, estimated it to be some 0·3% of the total chlorophyll; its contribution to absorption

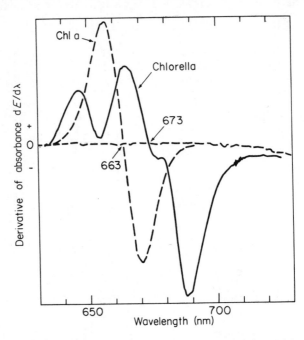

Figure 7.4. The derivative of the absorption spectrum of chlorophyll *a*, in ether solution (dashed curve), and in *Chlorella* (continuous curve). The derivatives cross the zero-line at the positions of the absorption maxima. The shoulder at about 680 nm indicates a second form of chlorophyll *a* in the alga. The dip at 654 nm represents the maximum absorption of chlorophyll *b*. From French, C. S. and H. S. Huang (1957). *Carnegie Inst. Washington, Year Book,* **56,** 267, with permission

or fluorescence phenomena must be negligible. The same applies to P690, which is a similar entity claimed by Witt's group to be the active centre for system II, on the basis of fast-reaction spectroscopy.

Several notations are in use to describe these chlorophyll forms, such as C_a670, Ca_{670}, $Ca670$ and so on. Since the only form in which chlorophyll *b* is observed is C*b*650, the *a* or *b* can be omitted. The classes C670 (or 673), C680 (or C683), C695 and C705 may be compared to the forms of bacteriochlorophyll observed in the purple bacteria: B800, B850 and B890 (see Figure 7.5). In these bacteria the analogue of P700 is P890 (to be distinguished from

B890). Bacteriochlorophyll absorbs at 770 nm in solution, chlorophyll at 663–670 nm in solutions, so in both cases the *in vivo* forms are displaced to the red. This may be brought about by aggregation of pigment molecules (it is known that such aggregates do have redder absorption maxima), or by increasing the hydrophobic nature of the environment (solutions in dry solvents absorb further to the red than when traces of water are present), or thirdly by complexing the pigment to protein or other non-pigment material. For the purposes of this section, in which we are concerned with structure of the thylakoid, we should note

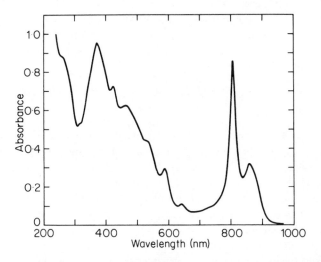

Figure 7.5. Absorption spectrum of *Rhodopseudomonas palustris* Van Niel strain 2137. Note the many peaks and shoulders, especially those at 804 nm, 858 nm and 880 nm. From Olson, J. M. and E. K. Stanton (1966). In L. P. Vernon and G. R. Seely (Eds.) *The Chlorophylls*, Academic Press, New York, p. 381, fig. 4, with permission. Copyright held by Academic Press

that the extent of these red-shifts, and that proportions of the various types, vary both in different species, and in different preparations. There may be reason to believe that the proportions of the 695–705 nm types, for example, may reflect more the kind of packing imposed on the chlorophyll by the conditions of the moment, and any alteration of the mechanical stress caused by disintegration, osmotic swelling or the forces due to the electric field believed to prevail during illumination, could alter the C695–705 material. This is the concept of 'plastic chlorophyll', which may also explain the absence of enhancement phenomena in some chloroplast preparations, as discussed in Chapter 6.

As mentioned above, at room temperature fluorescence of chloroplast material is mainly seen at 685 nm. Using the same notation, this is termed F685. At low temperatures (77°K) using liquid nitrogen, three components are usually visible, F685, F695 and F720–735. The latter is the most intense. F685 and F695 are evoked more by illumination of system II, while F720 is stimulated by both systems. Remembering that the dichroic chlorophyll was C695–705 (absorbing system I light) and that it fluoresced at 720 nm, the lack of specificity of F720 for systems I or II at 77°K may not be significant. If the molecules come close together, the likelihood of energy migration increases inversely as the sixth power of the distance.

The specificity of F685 for system II light is much more definite; at room temperature both F685 and photoreaction II are stimulated by light absorbed by an accessory pigment, chlorophyll b, or, much more strikingly, phycobilins in the red and blue-green algae.

We have to identify the fluorescing forms with the C670, C680, etc. The principal rule is that the value for the *Stokes' shift* (the difference between the absorption and fluorescence maxima) is of the order of 15 nm in most chlorophyll solutions. Applying this we have:

Absorption	C(b)650	C670	C680	C695–705
Fluorescence	nil	F685	F695	F710–730

Action spectra (measured in terms of oxygen evolution) show that system II is deficient in pigment absorbing at more than 700 nm, and has relatively less of C680 compared with C670. In addition chlorophyll b is almost entirely associated with system II, although this is not always clear-cut. We know that much of the C695–705 material is part of system I, which makes it a convenient energy trap from which P700 can draw its excitation. However if the low-temperature fluorescence observation really means that some F720 (hence C695–705) is attached to system II, first there must be very little, because of the red drop, and secondly it would be a better energy trap to feed P690 if it were say C695 rather than C705. It would not matter greatly if the energy trap was at a slightly lower energy than the sink (the photochemically active pigment), provided the difference could be made up by thermal energy. (It was suggested by Goedheer that C695 might act as a de-excitation pathway for system II in dangerously strong light, as an overflow.) Figure 7.6, based on a scheme of Govindjee, Papageorgiou and Rabinowitch, represents possible associations of chlorophyll forms with the two photoreactions. The diagram does not indicate the extent to which components, say C680 of system I, are in contact between different units, either of the same system or of system II. As drawn there is none, but there may well be some contact between systems I and II (the 'spillover hypothesis').

It might be added here that experiments by Goedheer on the ability of the carotenoid pigments to sensitize fluorescence of chlorophyll a indicate that in the red and blue-green algae β-carotene transfers its energy to system I; in green algae and higher plants it transfers energy to both systems I and II. The xanthophylls (oxygenated carotenes) do not appear to be active at all (but note the activity of fucoxanthin in Figure 6.2).

In conclusion, we have evidence of some organization of the pigments themselves, and of at least a degree of separation of chlorophyll and carotenoid pigments into specific entities associated with the two photoreactions.

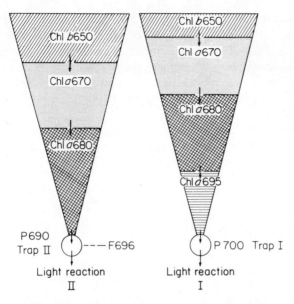

Figure 7.6. Diagram to illustrate the different distribution of the various forms of chlorophyll between the two photosystems. From Govindjee, G. Papageorgeou and E. Rabinowitch (1967). In G. G. Guilbault (Ed.), *Fluorescence*, Marcel Dekker Inc., New York, p. 511, with permission

7.3 Proteins

Trebst (1974) has presented an admirable summary of evidence which suggests that certain intermediates in the electron-transport chain have specific locations in the thylakoid membrane. Thus the acceptors of system I (ferredoxin, which is completely soluble, and the NADP-reductase) are on the basis of antibody and inhibitor accessibility assigned to the outer face, and the donors to system I on the inner face (plastocyanin and cytochrome *f*). The watersplitting system (the electron donor for system II) is commonly regarded as located on the inside face, but the evidence is indirect. The problem is that the watersplitting system fails if the thylakoid vesicle is opened; also kinetically it is harder to resolve the components of the pathway around system II than for system I. The chief impetus for these discussions is the support such differential location would give to the chemiosmotic hypothesis of energy conservation, which requires that electron transport shall generate either an electric field, or a pH gradient, or both, across the thylakoid membrane, without any prior formation of a 'high energy compound'. The resulting hypothesis appears in the widely reproduced figure by Trebst (see Figure 9.5). Connection between the two faces is supposed to be provided by the trans-membrane arrangement of the active centres, and the free diffusion of the lipid-soluble plastoquinone.

From a study of chloroplasts from species adapted to extremes of light intensity, Boardman and coworkers (1974) suggested that plastocyanin, cytochrome *f* and Y_{II}-components were also mobile, to the extent that they could form pools interacting with up to perhaps ten units of system I and a similar number of system II units. This was to explain the relatively constant efficiency of light energy conversion, but very different levels of light-saturation. The intermediates were relatively constant in proportion to each other, but the shade-chloroplasts possessed less of them in relation to Q and P700. Boardman's illustration is shown in Figure 7.7.

To some extent, the protein-based intermediates above can be regarded as extrinsic proteins, which, like the coupling factor, are embedded into or onto a more-or-less conventional bimolecular leaflet type of biological membrane. Even

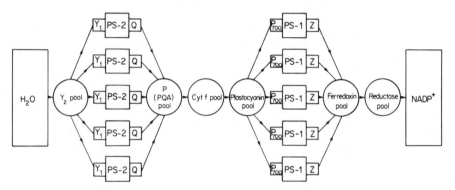

Figure 7.7. Schematic representation of light-harvesting assemblies, reaction centres and pools of electron carriers. Shade-adapted plants increase the number of photosynthetic units in relation to molecules of cytochrome *f*. From Boardman and coworkers (1974) with permission

the protein–chlorophyll *a* complex which operates the system I reaction centre may not penetrate right through the thylakoid (it is readily extracted by digitonin leaving intact membranes), and the chlorophyll (*a* + *b*) complex, which may provide 50% of the thylakoid protein, is now to be regarded as a binding agent, and hence clearly extrinsic.

Both the chlorophyll *a* complex and the (*a* + *b*) complex have been resolved into their component polypeptides. Some of these are easily recognized in preparations from unfractionated material (Anderson and Levine, 1974). Originally it was thought that the absence of system II could be inferred from the absence of the complex II–polypeptides, in stroma thylakoids, but this makes better sense if complex II is regarded as involved with granum formation, not specifically connected to system II.

7.4 The thylakoid membrane: sub-lamellar units

The thylakoid membrane, enclosing a space, is a necessary structure for photophosphorylation according to the chemiosmotic model (see Chapter 9). Witt and coworkers (1969a) has argued that the whole thylakoid membrane is the 'unit' for the ion movements, on which phosphorylation is held to depend, since the uncoupler gramicidin has a half-maximal effect at a molar proportion of 1×10^5 chlorophylls. Witt calculates that 10^5 chlorophyll molecules cover an area of $2 \cdot 5 \times 10^5$ nm^2, which is the order of size of a thylakoid. However, the quantities of redox carriers such as cytochrome f are such that each electron transport chain contains 600 chlorophylls or less, about 300 in each system. Hence breakage of the thylakoid down to a size at which it could no longer seal to form the necessary vesicle for phosphorylation might still allow it to show electron-transport activities.

Moderate breakage of the thylakoid does not impair any of its processes so long as the fragments can seal into ion-tight vesicles. Experiments with disintegration by sonic vibration or passage through a needle valve (the French press) followed by fractionation of the particles according to size in the ultracentrifuge have been used to identify the smallest particle that retains activity. Although various results have been reported, the more recent work (see Izawa and Good, 1965) indicates that in particles of some 2500 chlorophyll molecules (the size of the original CO_2-fixing 'photosynthetic unit' of Emerson and Arnold) the Hill reaction is seriously reduced in efficiency. The hope that by means such as this, individual active photosynthetic units could be isolated, has not yet been realized. However, it has been found possible to break the thylakoid into fragments and obtain at least a partial separation of system I and system II activities.

Michel and Michel-Wolwertz (1969) have achieved a degree of resolution by mechanical means, using a French press (see above) through which the chloroplasts were forced at a pressure of over 5 tons per square inch. They obtained three bands on a sucrose density-gradient (see Figure 7.8); the lightest band had a higher chlorophyll a/b ratio, and reduced NADP with an artificial electron donor (but less well than the original homogenate), while the other two bands had a reduced a/b ratio, and hardly reduced NADP at all. Bands 2 and 3 reduced DPIP in the Hill reaction almost twice as fast as the initial homogenate while band 1 showed very much less activity. Hence band 1 is enriched in system I and bands 2 and 3 in system II. The sizes of the particles in the bands were not determined, but there is a clear inference that there is a mechanical weakness between distinct system I- or system II-specific parts of the thylakoid.

Earlier, Boardman and Anderson (1964) achieved similar results, using the surface-active steroid digitonin. Differential centrifugation gave a series of fractions, the lightest of which were enhanced in the ascorbate–NADP (system I) reaction (plastocyanin was required). The heavier fractions, which appeared physically much larger than the smaller particles, were more active in the Hill reaction and had a reduced ratio of chlorophyll a/b. The separation does not work in low-salt media, nor with chloroplast lamellae that are relatively lacking the chlorophyll $(a + b)$–protein complex (below).

Application of more powerful detergents such as sodium dodecyl sulphate (SDS) (Sironval and coworkers, 1966) and sodium dodecylbenzene sulphonate (SDBS) (Thornber and coworkers, 1967a) produced particles small enough to be resolved by electrophoresis on polyacrylamide gels. One particle was obtained virtually devoid of chlorophyll *b*, and another with chlorophyll *a* and *b* approximately equal. The (*a* + *b*) band migrated the faster, except for a band of free pigment. Although these small particles were inactive photochemically, there was strong evidence that the *a* particle was associated with the system I preparations of Boardman and Anderson, and the (*a* + *b*) particle with system II. From their small size they were regarded as complexes of protein, chlorophyll and detergent, with some lipid, and other pigments. In their amino acid composition the protein components of the particles were different. The structural protein prepared by Criddle (1966) referred to earlier in this chapter had a uniform molecular weight of some 23 000, compared with values of 150 000 obtained by

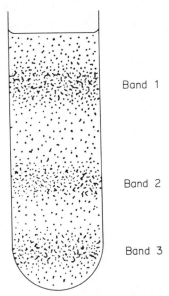

Band 1

Band 2

Band 3

Figure 7.8. The three-banded pattern of broken spinach chloroplasts centrifuged on a sucrose gradient at 60 000 g for 45 min. From Michel, J. M. and M. R. Michel-Wolwertz (1969) *Carnegie Inst. Washington Year Book,* **67,** 508, with permission

Thornber for chlorophyll *a*–protein–SDS complexes. However there could be a 25 000 m.w. subunit to explain the discrepancy. The general properties of the pigment-free proteins from both procedures were in the main similar.

The neutral detergent, Triton X-100, inactivates the Hill reaction and phosphorylation at the same rate (Deamer and Crofts, 1967), and the same range of concentrations causes changes in the electron-microscopic appearance of the thylakoid membranes. Surprisingly, the ascorbate—indophenol–NADP reaction after inactivation by low concentrations of Triton reappears at higher ones, and a particle has been prepared in 1% Triton by Vernon and coworkers (1966). This particle contains 1 P700 molecule per 200 chlorophylls, and the ESR signal attributed to P700 (Treharne and coworkers, 1963) was detected when the preparation was illuminated.

The preparation of chlorophyll–protein–detergent particles does not prove the existence of chlorophyll–protein complexes in the orginal thylakoid, since under these conditions chlorophyll could attach itself to a variety of proteins. However the remarkable activity in system I shown by Vernon's Triton X-100 preparation is a clear demonstration that such a protein–chlorophyll complex would provide a satisfactory basis for understanding system I. Furthermore, the author observed (Gregory, Raps and Bertsch, 1971) that the SDBS—protein–chlorophyll complex I, ascribed previously to system I, was absent from a mutant of *Scenedesmus* (mutant 8 of Bishop) which was previously known to lack the reaction centre of system I. There is hence a strong case for assuming that System I depends on the specific chlorophyll–protein complex I (and, by analogy, that a specific complex underlies System II also). However, there is good evidence that the second chlorophyll–protein complex prepared with SDBS (containing chlorophylls *a* and *b*) is not part of the photochemical apparatus: a mutant of barley obtained by Highkin and known to possess undetectable levels of chlorophyll *b* did not yield any of the complex (Thornber and Highkin, 1974). Deficiencies of the complex have also been observed in other mutants, and in every case there appears to be little or no formation of grana. It has been suggested that the $(a + b)$ complex has a major structural function in causing the thylakoid membranes to adhere together in the formation of a granum. This is in accord with work that has suggested a deficiency in the polypeptides of this complex obtained from stroma lamellae, separated from grana by French press treatment. There is however also often a deficiency of system II in some agranal chloroplasts or stroma lamellae owing to a block in the electron-transport chain at or near Y_{II}. In normal chloroplasts the $(a + b)$ complex may account for half the total chlorophyll, including virtually all the chlorophyll *b*, and it is clearly efficiently coupled to the rest of the pigment, so that it functions as a *light-harvesting complex*. Studies in the author's laboratory using circular dichroism indicate that both chlorophyll molecules in the complex are influenced strongly by an asymmetric field, probably by interaction with the protein. This is in contrast to the chlorophyll *a* complex where there is a strong indication of chlorophyll–chlorophyll interaction. In fact circular dichroism appears to be the quickest means of identifying the two complexes. They provide the principal circular dichroism signal of fragmented chloroplasts (in which the thylakoid–thylakoid signal referred to in the previous section is lost) (Scott, 1974). This is evidence that the complexes are genuine components of the native thylakoid.

The experiments above in which mechanical breakage of the thylakoid resulted in some degree of separation of the photoreactions suggests that the structural segregation in the membrane of the components of either photosystem may be nearly complete. One would expect that the system I and system II regions are particles with mechanically weak connections between them. With this in mind we turn in the next section to a consideration of the evidence offered by electron microscopists for the existence of a subunit structure for the thylakoid.

7.5 Thylakoid structure by electron microscopy

The arrangement of the thylakoid lamellae within the chloroplast envelope was illustrated clearly by the electron micrograph of Plate 1 (p. 12). Many workers have attempted to examine thylakoid material at higher resolution in order to discern the details of any underlying subunit structure, the existence of which is made very likely by the experiments described in the previous section. In addition to the sectioning method, fragments of single thylakoid membranes can be placed flat on a specimen grid and 'negative stained' by evaporating on to them a dilute solution of a salt of a heavy metal, which settles mainly in the crevices and hollows of the membranes, so that the surface pattern can be seen in relief in the electron microscope. Thirdly, use has been made of the recent 'freeze-etching' technique, in which a living leaf is rapidly frozen, and a piece of the resulting ice-block fractured in a vacuum; water sublimes from the fracture surface, leaving the non-volatile structures of the tissue standing out in relief. A replica of this surface can be made by depositing a thin layer of metal, and this is examined in the microscope, usually in conjunction with the 'shadowing' method in which heavy metal is deposited at an angle so that the height of the relief can be judged from the length of the shadows seen on the photographs. In this way structures are revealed without the use of fixatives or stains, and there is more confidence in the fidelity of the representation.

A considerable advance was made by Park and Pon (1961) using negatively-stained fragments of thylakoids obtained by sonic disintegration and differential centrifugation. There was a clear indication of a pattern, leading to the proposal that the membrane was made of a double layer of oval particles, 10 nm by 20 nm in dimensions, which were termed 'quantasomes'. The smallest particles of the sample (which were precipitated only by centrifugation for two hours at 144 000 g) appeared to contain some six quantasomes. The same sample was analysed, and the composition of the quantasome appeared in Table 7.1 (Lichtenthaler and Park, 1963). The assumption was made that the quantasomes accounted for all the material of the thylakoid, and that there was little or no cement or matrix between them. Preparations of these particles, which contain aggregates of various sizes, have been termed 'quantasome preparations' by several workers, which may have led to some confusion; suspensions of single quantasomes were never prepared.

Further confusion arose when it was found that there were several other kinds of particle attached to, or indented into, the membrane. For example, Sironval and others (1966) observed rows of 9 nm stalked spheres attached to the edges of membrane fragments and seen in profile with negative staining (Plate 9, p. 73). These particles, termed oxygenomes, were seen more clearly after digitonin treatment, and were found to cover, more or less, the surface of the thylakoid. Since then the identification with system II of photosynthesis has been discarded in favour of the view of Racker that they contain enzymes concerned with the formation of ATP (see Chapter 9). (The 9 nm particles can be isolated by washing the membrane with ethylenediaminetetraacetic acid, a chelating agent for metals such as Ca^{2+}.) In addition, the enzyme carboxydismutase also appears as particles about 10 nm in diameter, often in regions where it has an almost crystalline regularity of arrangement. The dimensions of the lattice in these regions is close to the figure given for the original quantasome pattern seen by Park and Pon, leading to the question whether the particles were attached one to each quantasome, or whether the quantasome was a confused appearance of such a regular region of carboxydismutase. This question was sidestepped when Branton and Park revised their view of thylakoid structure on the basis of their freeze-etching observations (see below).

Meanwhile, Kreutz (1966) had studied thylakoid preparations by means of low-angle X-ray scattering, and had proposed a model containing particles

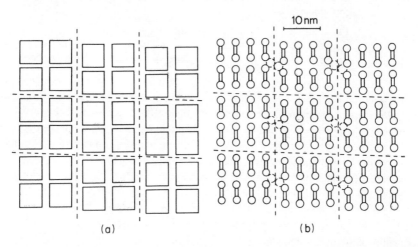

Figure 7.9. Diagrams by Kreutz combining his X-ray scattering results with the quantasome concept. The shapes are 'mass-centres' as seen from (a) the lipid side (outside) of the thylakoid, and (b) from the protein side (inside). (From Kreutz, W. (Ed.) *Biochemistry of Chloroplasts,* Vol. 1, Academic Press, London, p. 83, with permission)

arranged in squares of side 4·14 nm, grouped in larger rectangles 20 nm by 14 nm (see Figure 7.9). This agreed reasonably with the current estimate of the quantasome size (18 nm by 15 nm, Park and Biggins, 1964) but the question as to which particle was actually being observed still arose.

Weier and Benson (1966), on the basis of their micrographs of stained sections, proposed a different subunit model (Figure 7.10) which certainly appeared to relate to the substance of the membrane itself and not to any superficial material. However it is not easy to relate their model to those obtained by the other techniques, and there is also the problem of how far fixation and staining techniques can be trusted at this extreme degree of resolution. Unfortunately none of the other methods give a cross-sectional view.

A technique termed 'freeze-fracture' was developed, principally by Mülethaler (1966) and Park and Branton (1967), in which frozen specimens were cleaved by an advancing knife, and the fracture-surface (under vacuum) preserved by the deposition of a metal film ('replica'). This replica, after shadowing, showed

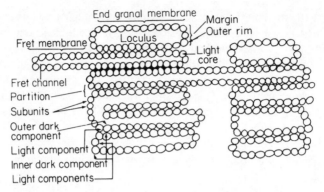

Figure 7.10. A hypothetical structure of thylakoids in section, from electron microscopy. From Weier, T. E. and A. A. Benson (1966). In T. W. Goodwin (Ed.) *Biochemistry of Chloroplasts* Vol. 1, Academic Press, London, P. 95, with permission

particle-laden surfaces. The original fracture surface could be further developed by allowing sublimation of ice to take place, so that freeze-etched regions were generated. After some debate, the view of Branton was accepted that the particle-laden surfaces were newly-formed by the splitting of a single membrane within its thickness, as when one separates the two sides of a piece of toast (Figure 7.11). The interpretation made by Branton and Park was set out in the first edition of this text; however, further study (see the review by Staehelin and coworkers (1976)) has led to an extended version, and to a new systematic nomenclature for the various surfaces. This was agreed by many of the workers in the field in a 14-author paper (Branton and coworkers, 1975).

Freeze-fracture terminology. Any surface derived from a biological membrane can be described with three binary parameters: P/E, F/S and u/s, so that eight kinds of surface may be expected (PFs, ESu, etc.). The P/E choice is based on the

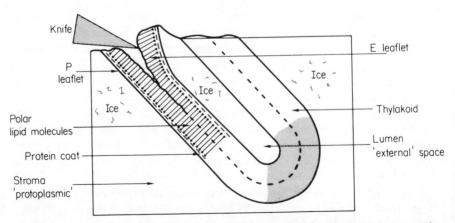

Figure 7.11. The bimolecular leaflet (protein–lipid) model for biological membranes, and the principle of freeze-fracture giving E and P fragments

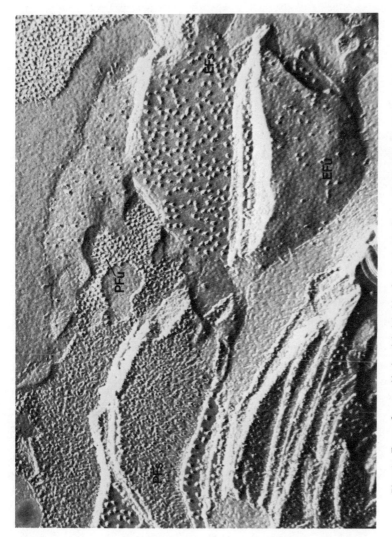

Plate 11. Freeze-fractured isolated thylakoids of spinach. The flat, partly-circular membranes of two grana stacks (left and right) appear connected by a stroma lamella. The typical appearance of the complementary pairs of faces PFs and EFs, and PFu and EFu are shown (see Figures 7.12 and 7.13). Note the aggregate of large (16 nm) particles on the EFs face. From Staehelin and coworkers (1976) with permission.

principle that all biological membranes separate two different liquid phases, one of which is Protoplasmic (such as the cytosol, the mitochondrial matrix or the chloroplast stroma) and one which is derived from the Extracellular medium (such as the endoplasmic-reticular spaces, vacuole contents, space between double envelopes and thylakoid lumina). Combining this principle with the bimolecular-leaflet concept of membrane structure, we can label one half P and the other E. This is shown in Figures 7.11 and 7.12. The asymmetry of membranes is important for many aspects of cell biology, such as electron transport, phosphorylation, hormone action and cell–cell recognition, and it is often insufficiently stressed.

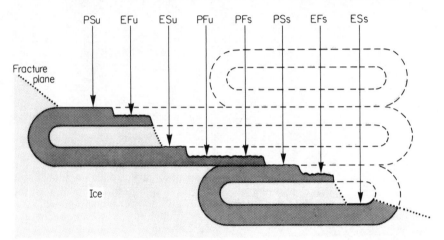

Figure 7.12. A thylakoid stack, freeze-fractured so as to show all eight possible types of face. (After L. A. Staehelin)

The F/S term distinguishes a fracture-surface (F), produced from within a membrane, from a pre-existing membrane surface (S).

The u/s term applies to thylakoids, which in the case of stroma lamellae and the outsides of the grana are unstacked (u), while inside the grana they are appressed (stacked, s). Note that the PSs surface can only be exposed by fracture, not etching. An unlikely display of all eight types is shown in Figure 7.12, and an illustrative micrograph in Plate 11.

Particle analysis. Staehelin analysed the distribution of particles of various sizes on the four fracture faces EFu, EFs, PSu and PSs. These are shown in Plate 11. Figure 7.13 represents the interpretation made. This work is elegant but complex, and for the reasoned connection between the many electron micrographs and the model the reader is referred to the review of Staehelin and coworkers (1976). The principal points of the model are (i) the EF particles (greater than 14 nm) project through or into the P-half of the membrane, (ii) both the EF and the PF particles (less than 14 nm) can combine with a third type of particle, which may be the light-harvesting chlorophyll (a + b)–protein complex, (iii) the small and large particles represent photosystems I and II respectively, (iv) the particles are accurately arranged so that, in a granum, the large particles on one thylakoid are intimately connected to the small ones on the next, (v) these particles, and also the external ones such as the coupling factor, are mobile and can redistribute themselves when changes in stacking take place, and (vi) there are fewer system II particles in the unstacked regions.

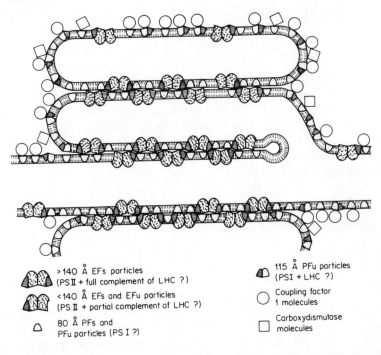

> 140 Å EFs particles
(PSII + full complement of LHC ?)

< 140 Å EFs and EFu particles
(PS II + partial complement of LHC ?)

80 Å PFs and
PFu particles (PS I ?)

115 Å PFu particles
(PSI + LHC ?)

Coupling factor
1 molecules

Carboxydismutase
molecules

Figure 7.13. Interpretation of the freeze-fracture and freeze-etching results. Note the distribution of the different types of particle between stacked and unstacked regions, the different degrees of membrane penetration, the associations of particles in stacked regions, and the binding in various amounts of the light-harvesting complex (LHC) to the two other particles. (From Staehelin and coworkers (1976), with permission)

Anderson (1975) concluded her review on the structure of the thylakoid by stressing the importance of these particles (the protein–chlorophyll complexes) and the problems posed by the unknown spatial relationship of them to the electron transport chains and to the coupling factors: 'All of these questions relating to molecular organisation present a challenge which has to be met before we can begin to understand the energy-transducing function of chloroplast thylakoids'.

CHAPTER 8

Photosynthetic electron transport

In this chapter we shall examine the evidence that certain redox materials in thylakoids and bacterial chromatophores are oxidized and reduced during photosynthesis, and the means by which such evidence is obtained. From there we can then discuss sequences of redox carriers (electron transport chains) at various key places. These partial sequences can be incorporated into more than one type of model, and some of the models currently proposed will be compared to the familiar zig-zag scheme.

8.1 Evidence for the participation of certain carriers in photosynthetic electron transport

8.11 Quinones

Plastoquinone (PQ) is the most abundant redox material in the chloroplast, its molar ratio to chlorophyll being $1 : 7$ of which some 80% is PQ-A, 20% PQ-C. This and its physical properties (see section 4.51) enabled quantitative extraction to be performed after various treatments so that changes in the redox state could be established. After illumination it was shown that a part of the PQ pool was reduced, although the reported degree of reduction varied rather widely. However it must be borne in mind that approximately half of the PQ of the chloroplast occurs in osmiophilic globules or plastoglobuli which may be metabolically inert and that these globules increase in size during the life of the leaf. It appears reasonable that most of the PQ of the thylakoid membrane can be reversibly oxidized and reduced during light–dark cycles.

Further evidence was obtained by Henninger and Crane (1967) that the Hill reaction with indophenol as oxidant was inhibited by exhaustive extraction of PQ with hydrocarbon solvents. Restoration of the activity required both PQ-A and also PQ-C in approximately their *in vivo* proportions. PQ-B may possibly act at the same site as PQ-C, but it is difficult to isolate and test.

Thirdly, spectroscopic investigations have revealed on illumination changes at 255 nm, which are indicative of quinones of this series (Klingenberg and coworkers, 1962); this method also allows determination of the kinetics, which is the necessary supplement to the foregoing evidence if it is to be decided whether the quinones are on a direct line of electron transport or a side-line. Thus, Witt, Suerra and Vater (1966) illuminated chloroplasts with 'system I' light (700–730 nm) followed by a flash of shorter wavelength. They observed a

255 nm change with a time constant* of 10^{-5} s, indicating the reduction of PQ by system II, followed by reversal in 10^{-2} s. While this last change is slow, it is consistent with the early kinetic experiments of Emerson and Arnold (1932) which showed that flashes of (neon) light were as effective as continuous light if the repetition rate was 10^{-2} s or faster.

Quinones can be reduced in one-electron stages, so that intermediate semiquinones are formed (p. 47). Spectra of plasto-semiquinone under various conditions were obtained by Bensasson and Land (1973), and one of these, the semiquinone anion, possessed an absorption band at 320 nm which corresponded with a 320 nm band appearing on illumination of system II preparations. It seems to be established, therefore, that plastoquinone is directly concerned with electron transport, that it is reduced by 'system II' light, and that PQ-A and PQ-C are both required.

Ubiquinone (UQ), which is the principal quinone taking part in electron transport in mitochondria, is found in the photosynthetic bacteria. Its role in photosynthesis is established by absorption changes at 270–275 nm, and by the demonstration that photophosphorylation is blocked in chromatophores from which UQ has been extracted, and is restored by the specific readdition of UQ. Plastoquinone is not found in the bacteria.

There is some variation in the isoprenoid side-chains of these quinones, but the commonest forms are PQ-9 in green plants and UQ-10 in bacteria.

Other quinones such as tocopheryl quinones occur with their reduced (tocopherol) forms, so that a similar function remains possible. Vitamin K_1 (a naphthoquinone derivative) has been shown to be less abundant in algae when they are grown non-photosynthetically, but there is no indication of its function.

8.12 Proteins

Plastocyanin is a protein of molecular weight 10 500 approximately, and it has been purified from several sources including higher plants as well as green, red and blue–green algae. It has not been found in the photosynthetic bacteria. Amino acid sequences are known for several species; there are 99 residues (98 in *Chlorella*). There is an intense blue colour caused by the single copper atom, although the broad absorption band centred at 597–598 nm ($\varepsilon = 4.5 \times 10^3$ l mol^{-1} cm^{-1}) makes recognition of absorption changes in the chloroplast very difficult, especially since the absorption by the pigments is much greater than in the 550–560 nm region where the cytochromes can be observed. It can be reduced by ascorbate and re-oxidized by ferricyanide ($E^0_{pH7} = 0.37$ V) and the differing degrees of absorption of the oxidized and reduced forms to the ion-exchange medium DEAE-cellulose made the preparation relatively straightforward (Katoh, Shiratori and Takamiya, 1962). Treatment of chloroplasts by ultrasound or with surface-active agents dissolves out plastocyanin, and the need

---

* The time constant of a reaction is an indication of its duration: for an exponential decay, the time constant is the time taken for 63% $(1 - 1/e)$ of the total change to occur.

to add this component back for certain reactions to take place is an indication of its importance. Further evidence for its involvement as an intermediate in photosynthetic electron transport was obtained by Gorman and Levine (1966) who showed that in a mutant of *Chlamydomonas* (Ac 208) which lacked plastocyanin, both photosystems were intact, but the connection between them was broken. The abundance is of the order of 1 molecule in 400 to 600 molecules of chlorophyll, which is consistent with an association with a single-quantum photosynthetic unit.

Cytochromes. The only *c*-type cytochrome characterized in chloroplasts of higher plants is cytochrome *f*, purified by Davenport and Hill (1952). The sharp α-absorption characteristic of reduced cytochromes was found at 554·5 nm (Figure 8.1) and work by Forti, Bertole and Zanetti (1965) established that there are four haem groups present in a molecular weight of 250 000 and that the molecule tends to disaggregate in solution. Davenport (1972) claimed to have separated a smaller haem-carrying protein. The abundance of cytochrome *f* in chloroplasts was estimated at 1 haem residue in 430 chlorophyll molecules (Davenport and Hill, 1952; Anderson Fork and Amesz, 1966). The redox potential at pH 7 (invariant) was found to be 0·36 V. The redox reactions of cytochrome *f* can be observed with relative ease *in situ* since the 'window' in the absorption of chlorophyll is well-placed, although the technique still calls for the measurement of spectra having magnitudes of the order of 10^{-2} absorption units.

In *Chlamydomonas reinhardii* the analogue of cytochrome *f* is cytochrome-553 (from its wavelength), and has been shown by Levine to have a role in electron transport adjacent to plastocyanin, using the 553-less mutant Ac 206. In *Euglena* the analogue is cytochrome-552. The Commission on Enzyme Nomenclature of the International Union of Biochemistry has introduced rules for naming cytochromes; in the absence of a chemical test the grouping of the last two with *c*-type cytochromes is speculative. They allocate the term c_6 (to replace *f*) in higher plants.

Table 8.1 sets out cytochromes discovered in two species of purple bacteria, from which it will be seen that the majority are classed as *c*-type. In fact the existence of a *b*-cytochrome in *Chromatium* was only recently discovered by Knaff and Buchanan (1975). In contrast to the bacteria, which have (as so far discovered) only one *b*-cytochrome each, green plants have only one *c*-type but several *b*-type cytochromes, although finding roles for the latter in the theory of photosynthesis is continuing to present great difficulties. However, Levine has observed that mutants of *Chlamydomonas* (Ac 115, Ac 141 and F-34; Levine, 1969) are blocked in the electron-transport system between the photosystems, and are deficient in cytochrome-559. This cytochrome is not necessarily homologous with 559-absorbing cytochromes in higher-plant chloroplasts, of which one exists in two forms, having redox potentials of 0·065 V and 0·37 V (see Bendall, Davenport and Hill, 1971). The high-potential form appears to be convertible into the low-form. Another 559-component has been claimed to be present in cytochrome-b_6 (563 nm), which was the first photosynthetic *b*-

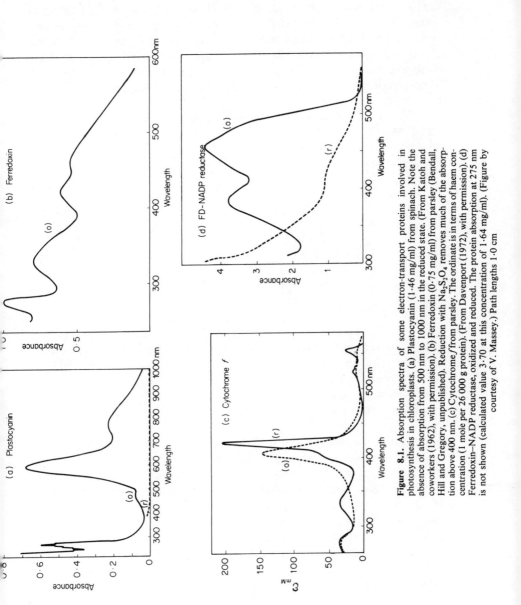

Figure 8.1. Absorption spectra of some electron-transport proteins involved in photosynthesis in chloroplasts. (a) Plastocyanin (1·46 mg/ml) from spinach. Note the absence of absorption from 500 nm to 1000 nm in the reduced state. (From Katoh and coworkers (1962), with permission). (b) Ferredoxin (0·75 mg/ml) from parsley (Bendall, Hill and Gregory, unpublished). Reduction with Na$_2$S$_2$O$_4$ removes much of the absorption above 400 nm. (c) Cytochrome f from parsley. The ordinate is in terms of haem concentration (1 mole per 26 000 g protein). (From Davenport (1972), with permission). (d) Ferredoxin—NADP reductase, oxidized and reduced. The protein absorption at 275 nm is not shown (calculated value 3·70 at this concentration of 1·64 mg/ml). (Figure by courtesy of V. Massey.) Path lengths 1·0 cm

cytochrome discovered (Davenport, 1952). The redox potential of cytochrome b_6 was found to be 0·00 V in chloroplast membrane material (Hill and Bendall, 1966). All the b-cytochromes mentioned appear to be lipoproteins and very tightly bound to the thylakoid membrane. However, by means of the detergent Triton X-100 and 4 M urea (which ruptures hydrogen bonds) soluble preparations of both cytochrome 559 (low potential) and b_6 have been obtained, as well as f (see, for example, Stuart and Wasserman (1975)). The properties of these preparations are similar in that each contained much lipid, and chlorophyll a; the protein part consisted of several small subunits. Both contained one haem

Table 8.1. Cytochromes in two purple bacteria

Cytochrome	Synonyms or absorption maxima	$E'_{0.7}$	
Chromatium			
C552	(C553, C423·5)	10 mV	possesses FMN
C555	(C556, C422)	340 mV	
cc'	(RHP, C430)	−5 mV	
b	560	−5 mV	
R. rubrum			
c_2	550, 415	338 mV	
cc' (RHP)	550, 423	−8 mV	
b	560, 430	0	

group which was identified as protohaem IX so that these are genuinely b-cytochromes. It was encouraging that the redox potential differed only slightly from the values expected from the determinations *in situ* lending credibility to both the determinations and the preparation method.

It should be pointed out that Lundegardh discovered another plant cytochrome absorbing at 559 nm in pea microsomes; this is known as b_3.

Ferredoxin-reducing substance and low-potential radicals. Trebst has reported experiments in which antibodies to thylakoid material were prepared in rabbits. The antigenic substances were found to be the ATPase 'coupling factor', the enzyme ferredoxin–NADP oxidoreductase, and an unknown component. The antibody blocked the reduction of ferredoxin by photoreaction I in chloroplasts. It was active against a soluble preparation obtained by San Pietro in a different laboratory, and this preparation would relieve the inhibition caused by the antibody. The preparation of San Pietro (ferredoxin-reducing substance, FRS) possessed heat-labile and heat-stable components. There was, however, no identifiable redox material, although at the time it was expected that some compound such as a pteridine might be found. While the results are not questioned, belief in this 'FRS' has subsided in favour of some more structural role for the preparation. Meanwhile Malkin and Bearden have pursued a signal obtained with electron spin resonance (ESR) (also termed electron paramagnetic resonance, EPR), a technique which detects the presence of molecules possessing one or more unpaired electrons (free radicals). Each orbital of a molecule can contain two electrons of opposite spin; when an orbital contains only one electron, it can be made to exchange energy with a microwave field by changing from one spin value to the other. The importance in this study is that any one-electron transfer process must either be detectable by changes in the number or kind of free radicals present. Note that transition metal ions with unpaired electrons, and molecular oxygen, are to be regarded as free radicals on this basis. ESR spectra have been identified for P700, the chlorophyll that forms the reaction centre of photosystem I, and at low temperatures (less than $25°K$) there is a signal resembling ferredoxin and termed 'bound ferredoxin' or 'bound ISP' (iron–sulphur protein) although no chemical analysis exists. A role can be argued for this radical as the acceptor (X_I) for system I. Hiyama and Ke (1972) reported a spectral change at 430 nm which they labelled P430; there was evidence that this was also a candidate for X_I.

Knaff and Arnon (1969a) discovered a spectral change at 550 nm labelled C550. This is unlikely to be a cytochrome and has since been suggested to be due to β-carotene, or even to phaeophytin a. It is closely related to the acceptor for system II (X_{II}).

Ferredoxin–NADP oxidoreductase, EC1.6.99.4. In non-cyclic electron transport in green plants there can be little doubt that ferredoxin is required for the reduction of NADP, and that the reduction of NADP by reduced ferredoxin is catalysed by the flavoprotein enzyme ferredoxin-NADP oxidoreductase, although it is possible that other flavoproteins may be present able to catalyse this reaction to some extent. The reductase has been isolated from spinach (Shin and coworkers, 1963). It has a molecular weight of 44 000 and contains 1 molecule of FAD. It is hard to establish a standard redox potential for flavoproteins in general, owing to the difficulty of defining the half-reduced state; in this case it is observed that at equilibrium the flavoprotein is extensively oxidized in the presence of extensively reduced NADP, and Massey and coworkers (1970) estimated a potential of approximately -0.38 V at pH 7. Two hydrogen atoms are carried by the single FAD molecule in the molecule. The spectrum (Figure 8.1) is altered by reduction, but there is a likelihood that the reduction proceeds in one-electron stages. Nevertheless the reduction of NADP is fully reversible in so much as the enzyme is an efficient catalyst of the diaphorase reaction, that is, the reduction of a dye such as DCPIP by NADPH. The protein is relatively strongly bound to the thylakoid, so that extensively fragmented material can usually reduce NADP in the presence of ferredoxin, but it is slowly lost on repeated washing.

Ferredoxin (see p. 61) was crystallized from parsley as a red protein catalysing the reduction by illuminated chloroplasts of methaemoglobin (Hill and Bendall, 1960a) in Cambridge, and simultaneously prepared by San Pietro and Lang (1958) at Yellow Springs, Ohio, who recognized its role in the reduction of NADP (then known as TPN: triphosphopyridinenucleotide), and it was named

PPNR (phosphopyridinenucleotide reductase). After the discovery of the NADP-reductase Arnon renamed PPNR 'ferredoxin', from its similarity to the iron–sulphur protein found in *Clostridium* (a non-photosynthetic obligately anaerobic bacterium) by Mortenson and coworkers (1962). An absorption spectrum of ferredoxin from parsley is given in Figure 8.1. The ferredoxins have been much studied as a class of protein; Hall argues that they must have been among the earliest proteins to appear in the evolutionary time scale, and has traced a 'family tree'. Amino acid sequences are known for the main types. They all show a characteristic ESR signal at $g = 1 \cdot 94$ at low temperature, in the reduced state. There are four types: type 1, from organisms such as *Clostridium*; 2, from green photosynthetic bacteria (which are obligate anaerobes also); 3, from purple photosynthetic bacteria (the best example here being *Chromatium*, an anaerobe) and 4, from plants. Type 1, it is argued, is a modified dimer of a 26-amino acid peptide (only half the usual number of amino acid types being represented). There are 8 iron and 8 labile-sulphur atoms possibly in two four-iron structures:

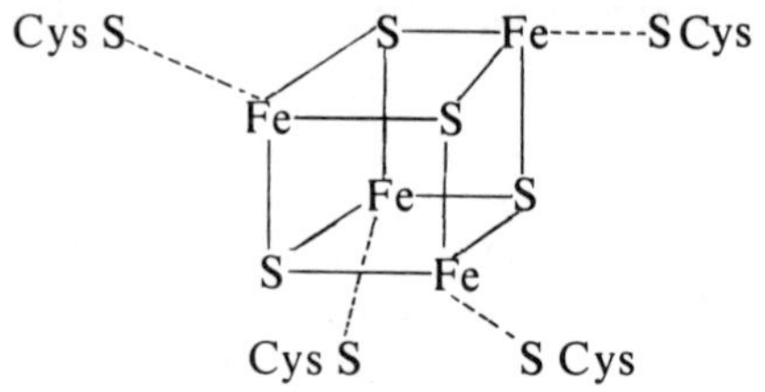

and the molecule carries electrons in pairs. Types 2 and 3 can transfer either single electrons or pairs; type 3 can be regarded as a trimer of the hypothetical 26-peptide. Type 4 has only 2 iron and 2 labile-sulphur atoms and transfers single electrons. The molecular weight is greater again, although it may be straining credulity too much to claim that the same hypothetical peptide is now represented four times. Nevertheless the sequence of ferredoxin types is a striking example of apparent gene-duplication and independent modification.

The iron–sulphide prosthetic group of all the types is held by cysteine–sulphur linkages to the iron, and is decomposed into ferrous salts and hydrogen sulphide by even mildly acid conditions.

The iron–sulphur proteins are an extensive class, including, for example, adrenodoxin in the adrenal gland of mammals, and other examples occur in mitochondria.

8.2 Sequences of redox carriers in electron transport in green plants

Given that there are two light reactions driving redox reactions, there must be a primary oxidant (X) and reductant (Y) for each. These symbols represent unknown intermediates that can be investigated kinetically. It is probably true to say that most of the work reported in this section has been done and presented on the basis of the series (zig-zag) formulation of the two-quanta hypothesis. A separate section will be given to assessing the merits of the zig-zag theory against others.

8.21 System I

The site X_1. Reference has been made (section 6.21) to the discovery of artificial redox couples with very negative standard potentials that could be reduced by chloroplast lamellae with system I light. By allowing the process to reach equilibrium, the actual potential of the electron source could be calculated from the observed redox level of the material added. Thus Zweig and Avron found a potential of approximately -0.5 V at pH 7·8, Kok reported -0.65 V, and Black -0.52 V. The primary donor is believed to be the chlorophyll known as P700 which is oxidized by light:

$$(P700)(X) \longrightarrow (P700^{+} \cdot)(X^{-})$$

Does the reduction of the viologen dye substitute for the reduction of X in the above scheme (in which case the reducing potential of P700 is given by the above measurement of -0.52 V) or is it reduced by X^- or a subsequent intermediate? The direct reduction by P700 can be discounted, since the speed of the reaction (measured by the rate at which the absorption of P700 declines) is very great (faster than 20 ns) and takes place at low temperatures, so that molecular movement cannot be part of the mechanism; the formation of $P700^+$ must be accompanied by the reduction of a non-diffusible X. Both the 'bound ferredoxin' and 'P430' have been suggested for this role. The correspondence between the rates of formation of $P700^+$ and reduction of P430 (see, for example, Ke, 1973), and the correspondence between the rates of disappearance of P430 and the reduction of soluble ferredoxin virtually force the conclusion that P430 is the primary acceptor, and that there is no other intermediate between P700 and the first chemically established and soluble product, reduced ferredoxin. The bound ferredoxin (or ISP) forms observed by ESR (see Table 8.2) has not been clearly identified with P430, since the ESR signal of the iron–sulphur proteins is only observable at temperatures close to absolute zero, making kinetic correlations difficult. By titrating the ESR signals with redox buffers, two species were found having potentials of -0.53 V and -0.58 V (Ke) or -0.563 V and -0.604 V (Evans). Although Ke considered that the more positive ISP could be provisionally identified with X and with P430, there has not been support for this and the two ISP forms were regarded by Evans and coworkers (1974) as secondary pools equilibrating with X. McIntosh and coworkers (1975) supported this view and claimed a new ESR signal 'outside the range of normal ferredoxins'.

Arnon, ten years ago, dismissed speculation concerning intermediates between P700 and (soluble) ferredoxin until 'a stronger electron carrier (meaning one with a more negative potential) is actually isolated from the photosynthetic apparatus'. While they have not exactly been isolated, a remarkable number of such carriers has now been observed, allowing full scope to creative theorizing. At the present time we have to sketch the sequence on the reducing side of system

Table 8.2. ESR signals from Chloroplasts

		g values	System	Assignment	References
i	'Signal I'	2·0025	I	P700	1, 2
			II*	P680*	3
ii		2·05, 1·94, 1·86	I	ISP† (centre A)	4, 5, 6
iii		2·05, 1·92, 1·89,	I	ISP†	4, 5
iv		2·06, 1·86, 1·76	I	X_1 (possibly ISP)	7
v	'Signal II'	2·0051	II	disputed	1, 8, 9
vi	'II'	2·00	II	possibly Y_{II}	9

The g value is a characteristic of an ESR signal, as wavelength is a characteristic of absorption spectra.
* Depending on the experimental conditions.
† ISP: iron–sulphur protein ('bound ferredoxin').
References: 1, Commoner and coworkers (1956); 2, Weaver and Weaver (1972); 3, Malkin and Bearden (1975); 4, Evans and coworkers (1974); 5, Warden and Bolton (1974); 6, Malkin and Bearden (1971); 7, Evans and coworkers (1975); 8, Esser (1974); 9, Babcock and Sauer (1975).

I according to Figure 8.2 in which X_1 is left unidentified, serving as a label with which P430 and the new ESR signal of McIntosh and coworkers (1975) may be bracketed. It is too early to dismiss the other ISP forms, which are shown as a branch, following Evans and coworkers (1974).

Continuing with the reducing sequence, the position of ferredoxin is virtually certain, as is the flavoprotein enzyme ferredoxin–NADP oxidoreductase (EC.1.6.99.4) that mediates the reduction of NADP. This sequence was well argued by Shin, Tagawa and Arnon (1963) who sorted out the relationships between ferredoxin (PPNR at the time) and two activities, a transhydrogenase (interconversion of NAD^+–NADPH and NADH–NADP+) isolated in the laboratory of San Pietro, and NADP (TPN)–diaphorase (reduction of a dye, in this case dichlorophenolindophenol, by NADPH) extensively purified by Avron and Jagendorf. Arnon's group crystallized the flavoprotein.

It may be noted that under conditions of iron-deficiency in blue–green algae ferredoxin levels may be low, and another flavoprotein, phytoflavin, has been observed to substitute for it (Bothe, 1969).

The site Y_1. The absorption band visible in chloroplasts at 700 nm, that is bleached either by illumination, or by oxidizing agents such as ferricyanide, is termed P700 and is believed to be a specific chlorophyll a molecule modified by its environment, although it has not been isolated (as, for example, a specific protein complex). The bleaching is presumed to be oxidation, and the calculated standard potential is 0·43 V (Kok, 1961). The oxidation is driven by system I light, and hence P700 is a candidate for the role of Y_1. Kinetically, the oxidation is fast (20 ns) and no component has been shown to be oxidized faster. The term 'chlorophyll a_1' is synonymous with P700.

There are three materials that, having potentials close to that of P700, may be examined as possible donors to P700. They are all proteins: plastocyanin

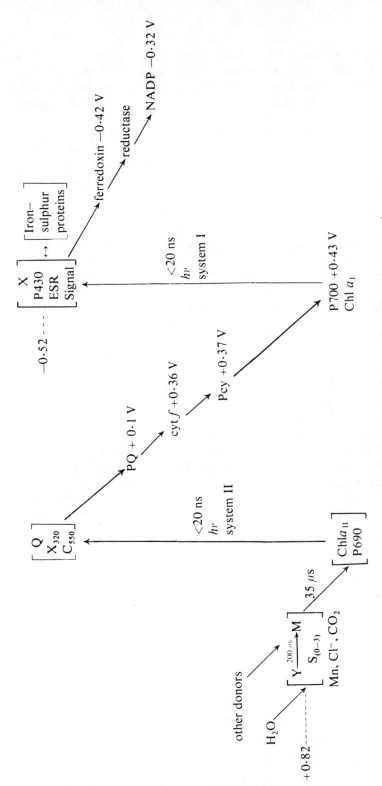

Figure 8.2. The zig-zag scheme of non-cyclic photosynthetic electron transport in green plants. Square-bracketed symbols indicate labels used in various laboratories for phenomena not yet identified with isolatable materials. The vertical scale indicates the standard redox potential (in neutral solution) of each intermediate, negative (i.e. higher energies with respect to the electron) in the upward direction

(0·37 V), and cytochromes f (0·36 V) and high-potential b-559 (0·37 V). The last can be discounted, since there are no spectral observations of its oxidation by system I, and also on fractionation of thylakoids with digitonin cytochrome b-559 is associated with the system II fraction, and the system I fraction is unimpaired (apart from loss of plastocyanin).

The original observation by Duysens and coworkers (1961) that f was rapidly oxidized by system I light and reduced by system II light (see section 6.12) was a demonstration of the existence of two light reactions, although it did not establish f as a direct intermediate. The algal varieties of f, c-553 and c-552, are essential in the sense that their absence blocks the intersystem electron transport in mutants of *Chlamydomonas* and *Euglena* respectively. However, there are certain differences between these cytochromes and higher-plant f, and also, with such complex lipoproteins, their absence might well produce crippling structural changes. The most satisfactory evidence comes from Witt's group; they measured the reaction time for the oxidation of reduced plastoquinone (at 275 nm) and found it to be equal to the reduction time of f, and the oxidation time of f was found to be similar to the reduction time of P700$^+$. This established f in a line of electron transport, although there have been several suggestions (such as that of Hill, 1965) that alternative pathways might exist, in parallel. Thus Haehnel (1974) 'titrated' the intersystem carriers by flashes of light, and found that more electrons were found on plastocyanin relative to f than would be expected from the relative abundance of each and from the close values of their potentials, and concluded that f was on a minority (10%) path, the principal carrier being plastocyanin. This view, while reputable, is opposed by the consensus alternative, that f and plastocyanin form a sequence. The main evidence for this is that the oxidation of f by P700$^+$ requires plastocyanin (contradicting reports from studies of system I particles may have underestimated the effect of the residual plastocyanin present). Plastocyanin itself is clearly needed for partial reactions of fragmented chloroplasts in which electrons from ascorbate enter system I via an artificial donor such as DAD, TMPD, or DCPIP (see Trebst, 1974).

DAD
(diaminodurene)

TMPD
(tetramethylphenylenediamine)

DCPIP
(dichlorophenolindophenol)

Plastocyanin has been shown to associate with, and react with chlorophyll a in a monomolecular layer at a water–nitrogen interface (Brody, 1975). This is argued to indicate a primary process thus

$$\text{Pcy} \cdot \text{Chl*} \longrightarrow \text{Pcy}^+ \cdot \text{Chl}^-$$

which is at variance with the usual scheme. We need to know the time course of the reaction, as it may be triplet-state dependent and, as such, too slow to be relevant. The observation is nonetheless of interest. There is less belief that plastocyanin may be structurally important; it is small, not a lipoprotein and readily soluble in water, so that the inhibition caused by its removal (or absence in mutants) and the restoration by its readdition may be more reliable evidence than is the case with *f*. Wood and Bendall (1975) have shown that the rate-constants for the reactions between *f* (*in situ*) and plastocyanin (in solution), and between plastocyanin (in solution) and P700 (*in situ*) are much greater than between *f* and P700. We may conclude that if *f*, plastocyanin and P700 are in a linear sequence, they are in that order, and plastocyanin occupies the position nearest to Y_1. The question of parallel paths, however, is in need of resolution.

8.22 System II

System II contains the mechanism for generating oxygen from water, a process which is biologically unique and chemically strange. It is the possession of this system that chiefly distinguishes green plants and blue–green algae from the photosynthetic bacteria. It is also the more labile of the two systems, in that most fractionating procedures that separate system I also destroy the oxygen-evolving apparatus. This lability lies in the actual watersplitting system, since if artificial donors other than water are used, particles can be prepared of relatively small size which carry out the photochemical reaction of system II. Even the digitonin procedure of Boardman and Anderson could not be said to genuinely 'prepare' system II particles, since the system II-active material was large and appeared to be the membranous residue. It seems that no oxygen-evolving system II preparation has yet been prepared that does not contain sufficiently large pieces of membrane for vesicle-formation to be possible, although the significance of vesicles in relation to oxygen production is not clear.

The site Y_{II}. Unlike the other sites considered, we are not in a position to select the more likely of a number of possible materials. No identified and chemically characterized materials are known which can be mapped on the donor side of system II. This is surprising since reference has already been made to cytochrome *b*-559 being oxidized by system II at 77°K, and Bendall, who discovered the high potential form, for a while suggested means by which it could contribute to the stepwise attainment of the high potential (0·82 V for 1 atm O_2 at pH 7) needed. However Cox and Bendall (1972) showed that the high-potential *b*-559 could be destroyed while leaving an appreciable part of the oxygen-evolving system intact. The oxidation of diphenylcarbazide (one of the artificial donors) was also found to be separable from high-potential *b*-559, as was the production of the signal P546 (=C550). They concluded that *b*-559 was not after all part of the direct electron-transport chain.

Other compounds such as the enzyme catalase (EC.1.11.1.6), which decomposes hydrogen peroxide to oxygen and water, might seem likely candidates for

Y_{II}, but are discounted first on the grounds that inhibitors such as the triazines have different effects on catalase and on the Hill reaction, and secondly that the production of hydrogen peroxide from water that would have to precede a catalase-catalysed step is energetically more difficult than the evolution of oxygen directly. (Cytochrome f has a remarkably vigorous catalase activity, which must be considered irrelevant.)

Manganese has been suggested as a possible Y_{II} component. Since the work of Pirson and of Kessler, it has been known that manganese-deficient algae were impaired in their photosynthetic activity, and it was later shown that this impairment lay entirely within system II. Kok and Cheniae (1966) have reviewed this topic; they consider that while there is no evidence of a valency-change in the manganese of the thylakoid during photosynthesis, nevertheless the Hill reaction does depend on the presence of strongly bound manganese. Protein is likely to be the ligand, but the valency state of the metal is not known. We may note that the high potential of the manganic–manganous ion couple (Mn^{3+}/MN^{2+}, $E_0' = 1 \cdot 5$ V, pH 7) offers an attractive basis for theorizing. Growth of algae under conditions of manganese deficiency causes them to lose their photosynthetic activity, and they live heterotrophically. The deficiency lies in system II. The chloroplasts of these algae are, however, still able to carry out a form of electron transport in which hydroxylamine (NH_2OH) acts as an electron donor to system II, given an acceptor such as ferricyanide. It seems as if hydroxylamine bypasses a manganese-requiring step. This is prevented by severe manganese deficiency, so possibly there is more than one site in system II where manganese is needed. The thylakoids of these deficient algae are grossly malformed, and this may provide part of the explanation; nevertheless it is observed that addition of manganese to a deficient culture causes photosynthetic ability to be regenerated in a matter of minutes, which might be thought to preclude protein synthesis or membrane formation.

Manganese is also removed by heat-treatment; incubation of chloroplasts at 45°C for 5 min severely diminishes manganese content and Hill activity,* although system I remains unchanged, and system II can be demonstrated both by its delayed fluorescence and by electron transport from donors such as diphenylcarbazide, hydroxylamine, hydrazine, hydroquinone, phenylenediamine or manganous (Mn^{2+}) ions. (Electron transport is easily detected by allowing system I to reduce methyl viologen, which takes up oxygen in the Mehler reaction, detectable with the oxygen electrode.) A similar inactivation of the oxygen evolving apparatus can be achieved by washing chloroplasts with $0 \cdot 8$ M tris buffer, but it is less certain that loss of manganese is the main factor in this case.

The chloride ion is also required for system II activity (Warburg and Lüttgens, 1944), and chloride deficient chloroplasts behave in the same way as those deficient in manganese, as has been shown by Izawa, Heath and Hind (1969). In

---

*The author was therefore considerably impressed by Dr. L. L. Tieszen's demonstration of millet (*Eleusine*) growing with a leaf temperature of 50°C or more (optimal) at Nairobi University. We have no explanation.

this case addition of chloride has an almost instantaneous restoring action. A third inorganic requirement for system II is carbon dioxide. This is quite independent of its metabolic fixation by stroma enzymes. It is possible that it acts, like manganese and the chloride ion, at site Y_{II}. (The requirement for carbon dioxide is fundamental to the theory of Warburg, see section 8.24, in which photosynthetic oxygen evolution is held to come directly from activated carbon dioxide with the one-step formation of carbohydrate.)

Some remarkable experiments by Joliot, Barbieri and Chabaud (1969) and by Kok, Forbush and McGloin (1970) indicate an electron storage system in the Y_{II} position. In one of the experiments the yield of oxygen in a bright short flash is

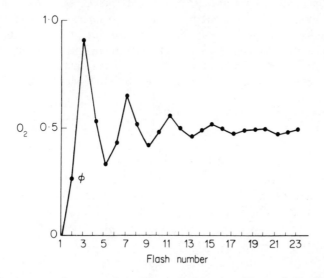

Figure 8.3. Amount of oxygen released by each of a series of flashes (of saturating intensity) from cells of *Chlorella* previously kept in the dark for 3 min. The dark interval between flashes was 300 ms. From Joliot, P., G. Barbieri and R. Chabaud (1969) *Photochem. Photobiol.* **10**, 309, Pergamon Press, with permission.

measured, then a second flash is given and the yield again determined. This is repeated some thirty times. Provided that the material had been kept in darkness for half an hour or longer, these workers were able to demonstrate that the yield of oxygen was periodic (Figure 8.3), being zero for the first flash and reaching maxima on the 3rd, 7th, 11th flashes and so on. This indicates clearly that there is a store of positive charges, filled one at a time by system II and emptied four at a time by the reaction

$$4OH^- \rightarrow O_2 + 4H^+ + 4e^-.$$

To account for the high yield on the third flash, Joliot suggested that partially filled sites could cooperate to form oxygen

$$|O| + |O| \rightarrow O_2$$

by a switch mechanism. Kok has suggested that if the store S passed through the sequence

$$S_0 \xrightarrow{h\nu} S_1^+ \xrightarrow{h\nu} S_2^{2+} \xrightarrow{h\nu} S_3^{3+} \xrightarrow{h\nu} S_4^{4+} \rightarrow S_0 + O_2$$

the phenomena could be explained if S_0 and S^+ were equally stable, even during long periods (30 minutes) of darkness. In Kok's view (which appears to be more attractive at the present time) each store functions independently of the others, and collects one electron at a time.

States S_2 and S_3 undergo deactivation, presumably by a back reaction between the store and one or more of the acceptors. One deactivation pathway may involve b-559 by system II at low temperatures, dependent on plastoquinone. This may or may not be the pathway of deactivation which is stimulated in the presence of a subgroup of the ionophorous uncoupling agents (see Chapter 9) named by Renger (1971) *ADRY* reagents: 'Acceleration of the Deactivation Reaction of Y'. Two of these are FCCP and Ant2s:

FCCP, carbonly cyanide p-trifluoromethoxyphenylhydrazone

$$F_3CO \langle \rangle NH \cdot N = C \begin{smallmatrix} CN \\ \\ CN \end{smallmatrix}$$

Ant2s (Büchel and Schäfer, 1970)

$$O_2N \langle \rangle \begin{smallmatrix} NO_2 \\ S \quad NH \end{smallmatrix} \langle \rangle \begin{smallmatrix} Cl \\ Cl \\ Cl \end{smallmatrix}$$

The mechanism of the ADRY effect is not clear since although charges appear to be lost from Y they have not been shown to appear anywhere else.

Another deactivation pathway involves the reversal of the photochemical reaction that produces light. This is the *delayed light emission*, or *delayed fluorescence*, of Strehler and Arnold (1951). The light is similar in wavelength to normal ('prompt') fluorescence, except that whereas fluorescence has a limetime of the order of 1 ns, and an intensity of the order of 1% of that of the incident light, the delayed fluorescence has an intensity very much less, and can be observed for several minutes. The actual intensity, as well as its decay-kinetics, vary greatly with the conditions of the experiment. Thus the traces obtained using the Becquerel phosphoroscope appear to show that there are two (or more) contributions, one that is fast ($0 \cdot 1$ to 1 ms) which is temperature insensitive and DCMU sensitive, and a slower component (20 to 200 ms) which is temperature sensitive and DCMU insensitive. These have been described by Bertsch (1969) and by other groups. However, Mayne (1969) and others have shown that there is an effect on delayed light emission of the pH gradient or the electric field believed to be connected with photophosphorylation (see Chapter 9). Both prompt and delayed fluorescence are apparently affected in the same way. Making corrections for the effect of the variation of the efficiency factor, Hipkins and Barber argued that there need be only a single component, which was temperature sensitive.

The dependence of the rate constant (k) on the temperature (T) is given by the equation

$$k = A \, e^{-E_A/RT}$$

where A is a constant related to the frequence of collisions, E_A is the activation energy and R the gas constant. By plotting log k against $1/T$, E_A was found to be of the order of $0 \cdot 6$ eV over a wide time interval. This value of $0 \cdot 6$ eV represents the difference between the energy of the red photon emitted ($1 \cdot 8$ eV) and the difference in the energy levels of the donor and acceptor in the reaction. We expect the donor and acceptor of the luminescent reaction to be close to the reaction centre chlorophyll, so that the measured value of $1 \cdot 2$ V represents the potential difference ($Y_{II} - X_{II}$). We also expect Y_{II} to be appreciably above the standard potential for the oxygen–water couple at pH 7 ($0 \cdot 82$ V). This would require X_{II} (=Q) to have a potential of some $-0 \cdot 3$ V, which was in fact observed by Cramer and Butler (see p. 150).

It would be expected that since the store S fills one electron at a time, there would be one or more detectable paramagnetic (unpaired electron) states giving ESR signals. In fact signals have been recorded from system II since 1956 (see Table 8.2), but the assignment of the possible components even to the donor or acceptor side is a matter of some dispute. The form of the spectra do not readily match plasto(semi)quinone, b-559 or manganese ions, although circumstances in the thylakoid could well be perturbing the expected signal. The reader is referred to the review by Joliot and Kok (1975) for further discussion of system II.

Spectral observations, and reaction-centre II. Work in Witt's laboratory has revealed a difference spectrum, with negative maxima at 687 nm and 433 nm, with respect to a control sample, brought about by illuminating an experimental sample with flashes of red light (610–710 nm) (keeping P700 oxidized with background illumination of 720 nm light). The magnitude of the difference was less than 10^{-3} absorbance units. These small changes, which are subject to very great noise levels, are detected with multiple flashes and averaging equipment; the signal to noise ratio improves as the square root of the number of observations made. The 687 nm change is termed chlorophyll a_{II} by Witt, and P690 in other laboratories by analogy with P700. In fragmented or damaged material the change appears at 685–683 nm ('P680'). The change decays with a variable half life. Two components of the decay have been resolved, one with a half life of 200 μs, and one of 35 μs. These are interpreted as successive electron donations from the watersplitting system. These times may be compared with the reduction time of X-320 (supposed to be plasto-semiquinone at X_{II}) which is complete in less than 10 μs. It is argued that the reduction of X-320 is considerably faster than that, since an electric field (detected by an absorption change at 520 nm, see Chapter 9), supposedly due to the primary charge separation, appears after system-II illumination in less than 20 ns.

In section 2.22 several types of photochemical reaction mechanism were described. The type termed photoionization applies clearly to system I, where the chlorophyll known as P700 is in close association with the acceptor X_I, so that the excited chlorophyll can cause the electron movement giving X^- and $P700^+$. By analogy, it is attractive to consider P690 in the same way. However, the 687 nm difference spectrum cannot be brought about by the application of oxidizing agents (reasonably so if the potential is so high) and therefore the peak is not proved to be oxidized chlorophyll. Döring (1975) has discussed the possibilities of chlorophyll a_{II} being a photosensitization reaction. This would involve a triple complex (chlorophyll–Y_{II}–X_{II}); the excited chlorophyll would cause the formation of Y^+ and X^- but no change in the chlorophyll. Döring could find no evidence to settle the question but pointed out that the photoionization mechanism provided at least a consistent rationale for observations.

Wolff and coworkers (1974) summarized the results from Witt's laboratory, setting out the kinetics of the reactions around system II, which are incorporated in Figure 8.2.

The site X_{II}. It is generally held that the level of fluorescence of chloroplasts, which is largely generated by system II, is controlled by the redox state of the

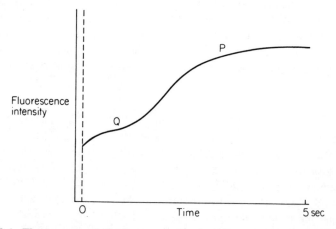

Figure 8.4. Time-course of fluorescence observed during illumination of isolated chloroplasts with short-wave light, after a period of darkness, or after an exposure to far-red light. The successive removal of two distinct pools of quenching material (Q and P) can be seen clearly. From Kok, B and Cheniae, G. M. (1966). In D. R. Sanadi (Ed.) *Current Topics in Bioenergetics*, Vol. 1. Academic Press, New York, p. 1, with permission

primary acceptor, Q, the quencher (see also Chapter 6). Within a few seconds of illumination (see Figure 8.4) the fluorescence increases from a low level to a maximum value, corresponding to the reduction of the primary donor.

Moreover, the steps in this *fluorescence induction curve* marked Q and P indicate the contribution of P which buffers the redox level of Q over a certain range. The inhibitor DCMU (diuron) inhibits system II by blocking the reaction between Q and P.

$$Cl \underset{}{\left\langle \begin{array}{c} Cl \\ \\ \end{array} \right\rangle} —NH—CO—N(CH_3)_2$$

Diuron or DCMU (3(3,4-dichlorophenyl)-1,1-dimethyl urea)

We wish to consider chemical candidates for the positions of Q and P. The curve of Figure 8.4 can be used to estimate the sizes of the Q and P pools, giving values for the total of between 1/8 and 1/30 of the chlorophyll present (on a molar basis). It is clear that Q cannot account for more than half of this and probably represents less. Nevertheless the only redox materials available in these proportions are the plastoquinones, of which PQ-A and PQ-C occur at one mole per 7 and 30 moles of chlorophyll respectively. However, because of the large and variable proportion of these quinones that are stored in the plastoglobuli it is difficult to take this calculation further. The potentials of the two were estimated by Carrier (1967) as 0·115 V and 0·055 V respectively (in absolute ethanol), but the potentials have not been estimated *in situ*. Other materials having potentials in this range are the cytochromes b_6 and b-559 (low potential) which are present in much smaller quantities. Although b_6 was originally placed at X_{II} it is associated with system I both in fractionation experiments and in its reactions. Cytochrome b-559 presents more of a problem since it is certainly associated with system II, and a cytochrome-559 is implicated in photosynthesis in *Chlamydomonas* from Levine's study of mutants. However, in higher plants the role of b-559, either high or low potential, is far from clear, but very unlikely to be concerned in the direct sequence of electron transport, since the spectral changes to be expected are not observed. There are however two spectral entities which are claimed to be associated with the primary acceptor, C550 and X-320.

Knaff and Arnon (1969a) published the results of experiments in which chloroplasts, in the presence of high concentrations of ferricyanide which prevented any cytochrome changes, showed a decrease in absorption at 550 nm on illumination. Both the wavelength dependence of the illumination necessary for this observation and its distribution after digitonin fractionation showed that it was a system II reaction, and the persistence of the effect at 77°K placed it close to the reducing side of the reaction centre. The effect was ascribed to a new component, and labelled C550. Studies in the laboratories of Bendall, Butler and Malkin established that plastoquinone and β-carotene were required for the appearance of C550 (these compounds were extracted by heptane, and were re-added), and that there was a reasonable similarity, in the kinetics of reduction and oxidation, between the two phenomena C550 and Q including the observation that DCMU did not inhibit the reduction of either. The reduction of C550 at 77°K appeared to be associated with the oxidation of cytochrome b-559.

Meanwhile Stiehl and Witt (1968) had found an increase in absorption at

320 nm, and further work in Witt's laboratory established by arguments similar to those used with C550 above that it represented a material labelled X-320 with the following properties: it was reduced by system II, the reduction being insensitive to both DCMU and a temperature of 50°C. The reduction was fast (20 ns), and the oxidation time matched that of Q (6×10^{-4} s). They argued that X-320 was the primary acceptor of system II and could be identified with Q.

By means of pulse radiolysis, Bensasson and Land (1973) obtained difference spectra for the semiquinone of plastoquinone, relative to the unchanged plastoquinone. Under conditions in which the semiquinone anion was formed (formula on p. 47), the difference spectrum contained peaks at 320 nm and 440 nm. The agreement with the X-320 spectrum is good and, for our present purposes, we can conclude that the site X_{II} is occupied by some environmentally-selected form of plastoquinone that is rapidly reduced by one electron to form the semiquinone anion. This is reoxidized when two reducing equivalents are transferred to plastoquinone, leading to the accumulation of fully reduced (plastoquinol) molecules. (There is no indication of the relative importance of the plastoquinones A and C, nor of the minor quinone forms known to be present.) Furthermore, this identification accounts for the requirement for plastoquinone in the C550 phenomenon, and there is a degree of agreement that C550 may be caused by a spectral shift of β-carotene brought about by the charge separation involved in the formation of the semiquinone anion.

It is however not clear at the present time what the relation is between the reactions of plastoquinone and the redox potentials of X_{II} obtained by Kok (see Table 6.1) and Cramer and Butler (1969), whose titration of relative fluorescence intensities with redox reagents yielded two mid-point potentials, in the ranges -0.02 V to -0.035 V and -0.27 V to -0.32 V. At the time the more positive potential was equated to that of X_{II} (Q) and taken to be a better estimate of the potential that Kok had measured indirectly. However, given a two-stage reduction of plastoquinone, the more negative potential might represent the formation of the semiquinone.

The reaction between plastoquinone (reduced) and oxidized cytochrome f has been mentioned above; it is well established as the slowest of the redox reactions and provides a rationale for the slow time of 10^{-2} s for the utilization of light energy measured by Emerson and Arnold. The reaction is pH-sensitive and has been used in investigations of the mechanism of coupling between electron transport and ATP formation.

In summary, the non-cyclic system of electron transport in chloroplast photosynthesis may be represented by Figure 8.2.

8.23 Cyclic electron flow

Certain materials, added to chloroplasts to serve as electron acceptors in the Hill reaction, were found to stimulate the production of ATP as expected, but the amount of ATP formed was indefinitely large in relation to the quantity of the material added. Since in the absence of metabolic enzymes the only source of

ATP known was that associated with electron transport (but see the discussion on *Halobacterium* in Chapter 9) it was assumed that these reagents catalyzed a cyclic process of electron transport around one or more photoreactions. This assumption remains reputable in spite of continuing difficulties in describing the pathway of cyclic electron flow and in naming the intermediates concerned. *Pseudocyclic* electron flow involves the whole chain and is observed when a reagent such as FMN (Figure 4.5) is employed. FMN is reduced by system I and reoxidized by molecular oxygen. The immediate product of the oxidation is hydrogen peroxide, but this is usually decomposed to water and oxygen by catalase. The level of oxygen is maintained by system II, so that no change in the quantity or redox state of oxygen or FMN can be observed, only the steady production of ATP by the concealed electron transport. This may however be immediately detected by its sensitivity to the inhibitor DCMU, its requirement for system II light, and its inhibition by the removal of oxygen from the system to begin with.

The cyclic photophosphorylation with PMS (phenazine methosulphate) or pyrocyanine (which converts readily into PMS) can be explained as a system I dependent process in which PMS is reduced by X_I and diffuses to Y_I or some point close to Y_I; we can rule out any involvement of plastoquinone, since the inhibitor DBMIB (see p. 168) is known to block plastoquinone, and has no effect

DBMIB: dibromothymoquinone, or 2,5-dibromo-3-methyl-6-isopropyl-*p*-benzoquinone.

here. In the same way the plastocyanin antagonist KCN fails to inhibit, leaving only P700 (or cytochrome *f* acting independently of plastocyanin). On the other hand DAD or DCPIP (section 8.21), when acting as cyclic cofactors, are sensitive to KCN, so that either cytochrome *f* or plastocyanin is the point of reinjection of electrons. This is sketched on the basis of the zig-zag scheme in Figure 8.5. The mechanism of ATP generation is left to Chapter 9.

At the time of the discovery of the above cyclic processes there was considerable interest since they offered a means for supplying extra ATP. The carbon fixation system, employing the reductive pentose cycle, was known to require $1\frac{1}{2}$ ATP molecules per NADPH, and the non-cyclic electron transport did not appear to provide more than 1 : 1. The requirement is in fact even higher if sucrose or starch formation is allowed for. Although non-cyclic electron transport may by itself generate sufficient ATP (see the discussion in Chapter 9), the researches of MacRobbie and Kandler (see p. 178) do show that ATP is made available from the chloroplast to the cell when non-cyclic electron transport is not possible. The cyclic system is expected to meet the shortfall.

There was assumed to be a 'natural' cyclic cofactor present, and from Arnon's laboratory it was claimed that ferredoxin was such a natural cofactor. Arnon and coworkers (1965) wrote a scheme in which cytochrome f was reduced by b_6, which in turn was reduced by ferredoxin at the X position. For other reasons he also held that there was no connection between this pathway and the non-cyclic pathway, a suggestion which has not met with much acceptance. Nevertheless the involvement of b_6 with f was reasonable since b_6 is associated with f in

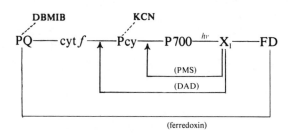

Figure 8.5. Electron-transport loops around system I, in the presence of ferredoxin, PMS or DAD as additives, which provide an explanation of cyclic photophosphorylation and its sensitivity to inhibitors. For its context this figure should be referred to the zig-zag scheme, Figure 8.2

digitonin-fractionated system I preparations, and in the more subdivided fractions obtained by Wessels and Voorn (1972). Furthermore there were observations (see, for example, Hind and Olson (1967) and Witt and coworkers (1969)) that b_6 was reducible by system I light. It has to be stated that b_6 is an ambiguous material at present, since other observations made b_6 also reducible by system II, which could be used to support a scheme in which f is bypassed by b_6. There is very little firm evidence that can take this discussion further.

Forti and Zanetti (1967) announced that the ferredoxin–NADP reductase could form a complex with cytochrome f, and would reduce f in the absence of NADP$^+$. This offered not only a neat cyclic pathway but also indicated how it could be regulated by circumstances: when metabolism became ATP-deficient, NADPH could accumulate; when NADP$^+$ had disappeared, electrons would return to f, generating ATP until the reductive pentose pathway had consumed the excess NADPH. They showed that an antibody to the reductase inhibited cyclic photophosphorylation.

Meanwhile in Trebst's laboratory a study of electron transport was made from the point of view of the location of the intermediates on one or other sides of the thylakoid membrane. This led to a scheme (see section 9.22) which provided an account of photophosphorylation in both the normal non-cyclic chain and in partial reactions around either photosystem, using artificial electron donors or acceptors. The scheme was readily adaptable to explain cyclic electron flow with PMS, and also with ferredoxin; ferredoxin unlike PMS is not lipid soluble and was expected to reduce an acceptor on the outside of the thylakoid, (the same side as the donor which reduced the ferredoxin) unlike PMS which passed through the membrane. The discovery that ferredoxin-catalysed cyclic electron flow was inhibited by DBMIB located the acceptor close to X_{II}, and Q was ruled out by the insensitivity to DCMU. Hence plastoquinone is the likely acceptor (Figure 8.5). This scheme of course contradicts Forti's hypothesis of the involvement of a complex between the NADP-reductase and cytochrome f; it also makes no use of his observation that an antibody to the reductase inhibits cyclic phosphorylation, unless plastoquinone can diffuse to the reductase inside the thickness of the membrane.

8.24 Comments on the validity of the zig-zag scheme compared with the others

The Z-scheme commands almost universal acceptance, in spite of the uncertainties that have been described. Criticism of it is difficult since many

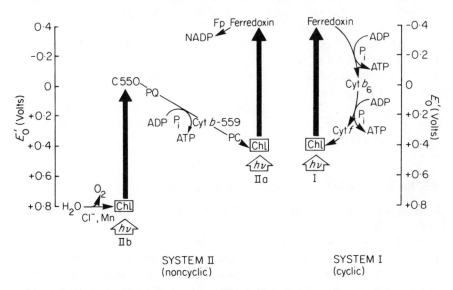

Figure 8.6. Scheme of Knaff and Arnon (1969b; with permission) for a three-light-reaction scheme in which system II is a zig-zag of reactions IIb and IIa. System I is reserved for cyclic electron flow

experiments are planned and interpreted on the basis that the Z-scheme is correct, and cannot be usefully read on any other basis. Perhaps the most interesting, and the most recent, alternative is that of Knaff and Arnon (1969b), which is represented in Figure 8.6. This contains the same isolated cyclic photosystem 'system I' as Arnon and coworkers (1965), with a two-light-reaction version of system II making use of the then newly-discovered C550 and high-potential b-559. Both light reactions of system II have to have the same wavelength dependence, however, so that the scheme failed to satisfy the observations concerned with light of varying wavelengths quoted in section 6.1.

A theory has been proposed by Franck and Rosenberg in which there is only one reaction centre, but it receives both singlet and triplet excitation collection by the pigment systems. This model can explain the enhancement phenomena. The reaction centre is supposed to act twice on a redox carrier such as cytochrome f (which represents X_{II} and Y_I) and generates the oxidant and reductant (Y_{II} and X_I) which produce oxygen and reduce NADP. The existence of the other redox materials with intermediate potentials is not in accord with the theory, and the fractionation of the thylakoid into I and II particles is a further difficulty.

Warburg's hypothesis, in which light splits activated carbon dioxide directly to carbohydrate and oxygen, three-quarters of which then recombine as in respiration producing the energy for the activation of the carbon dioxide, demands the complete reversal of the foregoing concepts of electron transport, in favour of pathways not yet discovered. The hypothesis is based experimentally on an apparent quantum requirement of four per carbon dioxide molecule fixed, on the (agreed) requirement for carbon dioxide for maximum Hill reaction activity, and of a large store of loosely bound carbon dioxide in algae. A summary of Warburg's position can be found in Warburg and coworkers (1969). One might leave this scheme as an exercise in model criticism, or acknowledge, following Hill (1965), that it has played a most useful part in stimulating interest, experiment and discussion in this field.

It is hard to escape the conclusion from this survey that the zig-zag model presents more satisfactory explanations for a greater proportion of the observations than any of its rivals.

8.3 Photosynthetic Electron Transport in Bacteria

Whereas in the green plant the most obvious electron transport system is non-cyclic, it has been argued that most of the photosynthetic electron transport in bacteria is cyclic. This may seem unlikely since photosynthesis in both groups reduces carbon compounds and oxidizes an environmental hydrogen donor. However the important difference is that in bacteria the hydrogen donors are of low potential and can be expected to reduce metabolic coenzymes without any need for light energy. Water on the other hand would not be thought of as a hydrogen donor without a considerable energy supply.

Bacteria may be thus able to reduce NAD without the use of light. Therefore the only *need* for light in reductive metabolism is to provide ATP. This can be achieved by a cyclic system. It is not clear whether materials of more reducing potential than NADH (particularly ferredoxin) can be reduced by such hydrogen donors as inorganic sulphur compounds or simple organic molecules without light energy. Certainly in the *Clostridia* pyruvate can be oxidized to acetyl phosphate and CO_2 with the reduction of ferredoxin (the phosphoroclastic split). If not, then bacteria carrying out the reductive citrate cycle would need, apparently, non-cyclic electron transport to reduce ferredoxin. However, the phenomenon of reverse electron transport in mitochondria (discussed in Chapter 9) allows NAD to be reduced by succinate against a standard potential difference of some 0·3 V, using energy of hydrolysis of ATP. Hence it is possible (but not at present established) that ferredoxin could be reduced by means of NADH, using ATP from cyclic electron transport.

Non-cyclic electron transport can however be demonstrated in isolated chromatophores of for example *Rhodospirillum rubrum*. Ascorbate donates electrons (mediated via the dye dichlorophenolindophenol) to NAD in a light dependent reaction. The reaction is inhibited by low concentrations of 2-*n*-heptyl-4-hydroxyquinoline-*N*-oxide (HOQNO) but not by antimycin A. On the other hand the reduction of NAD by succinate, which depends on ATP from a presumed cyclic system, was inhibited by both reagents at low concentration (Sybesma and coworkers, 1972).

Primary processes. Just as with P700 in green plants, so absorption changes have been recorded in bacteria in the fraction of pigment absorbing furthest to the red, brought about either by illumination or by application of oxidizing agents. Although the pigment concerned is bacteriochlorophyll *a* in all but one species, the wavelength of absorption of the photoactive fraction varies, being for example 870 nm (P870) in the purple non-sulphur bacteria *Rhodospirillum rubrum* and *Rhodopseudomonas capsulata*, 890 nm (P890) in the purple sulphur species *Chromatium D* and 840 nm (P840) in the green sulphur genus *Chlorobium*. In *Rps. viridis*, where bacteriochlorophyll *b* is present, P960 is observed. Frenkel (1970), in a useful and comprehensive review, introduced the notation P_a to indicate any of these photoactive entities. Redox potentials of these presumed reaction centres have been reported between 0·44 V (*R. rubrum*), 0·49 V in *Chromatium* and 0·525 V in *Rps. spheroides*. Parson (1968) showed that in *Chromatium* P890$^+$ appeared in less than 0·5 μs, and few would object to the working hypothesis that the oxidation of the photoactive bacteriochlorophylls was a primary process, homologous with the oxidation of P700 in green plants.

As with green plants, there are difficulties over the primary acceptor, X. On the one hand, work by Slooten (1972) established a spectral change accompanying P870$^+$ formation in Athiorhodaceae, which was shown by Bensasson and Land to be very similar to their spectra for the difference spectra (semiquinone-anion minus quinone) of ubiquinone. Supporting this, it was observed that reaction centre preparations were active in the P870 reaction while containing 3 moles ubiquinone per mole of P870, and less than 0·02 moles of *b*-cytochrome. Evans and Crofts (1974) titrated the oxidation of P870 (at a subsidiary wavelength of 605 nm) and obtained a potential for the acceptor of −0·025 V; studies of fluorescence in the presence of reducing agents (see Clayton, 1971) also showed a primary acceptor with a potential of this order. A much lower potential was measured by Govindjee and coworkers (1974) of −0·37 V, although, as argued with respect to green plants, from the titration of fluorescence of Cramer and Butler (1969), such low potentials are not inconsistent with a primary acceptor being the quinone–semiquinone couple. The higher potential of approximately zero volts would apply to the quinone–quinol couple and would be consistent with the data. Evidence against ubiquinone being involved at X comes from the application of dibromothymoquinone (DBMIB) which is a competitive inhibitor of quinones. Baltscheffsky (1974) found that the reduction of the b-cytochrome of $R. rubrum$ was not inhibited, but its reoxidation was, and he placed ubiquinone in the pathway between b and cytochrome c_2, making no suggestions as to the nature of X.

DCMU is not active in bacteria; since it appears to inhibit primarily at the X_{II} position (argued above to be plasto-semiquinone) in green plants, by analogy in bacteria X ought to be different.

An absorbance change has been detected at 446 nm (P450) in bacteria, which may be analogous to P430 in plants. ESR signals have been detected both for P_a^+ ($g = 2.0026$) and for other transients, but there is difficulty in the kinetic analyses. X^- should have a rise time equal to that of P_a^+, and both the optical and ESR kinetics should agree. An interesting discussion of these difficulties is given by Wraight and coworkers (1975) with respect to $Rps. sphaeroides$. A role for both quinone and iron (non-haem) is argued in the review by Parson and Cogdell (1975).

Secondary electron transport. The appearance of P_a^+ is followed by the oxidation of a c-cytochrome. There are in most cases more than one c-type present, and the relationships are not clear. In several species it has been suggested that there are multiple pathways (see the review of Frenkel, 1970) and that each has its specific c. Evidence for this is the finding by Parson (1968) that the rate of reduction of P890$^+$ is matched by the rate of oxidation of c-555 in $Chromatium$. This implies a sequence. However, the reaction does not take place at very low temperatures, when the much slower oxidation of c-552 is then observed, so that c-552 must be very close to P_a. This may be a divergence from the same reaction centre, or each pathway may have its own centre. The latter hypothesis was strongly favoured when Morita (1968) found that the relative contributions from the bacteriochlorophyll types B800, B850 and B890 to the oxidations of the two cytochromes were different, as if each reaction centre had a specific action spectrum. This argument was weakened when Parson and Case (1970) showed that the effect only appeared under conditions when the efficiency of excitation transfer could be expected to vary (owing to such forces as the fields connected with phosphorylation; see Chapter 9 for discussion of the concept of plastic chlorophyll). Better evidence came from the kinetic experiments of Sybesma and Kok (1969), for the existence of two light reactions in $R. rubrum$ for cyclic and non-cyclic electron transport.

It seems to be a reasonable generalization from the reported examples that oxidized c-cytochrome is reduced by a b-type, although not necessarily directly. That uncertainty is related to the uncertainty in allocating X, above. It was believed that b-type cytochromes were only found in the Athiorhodaceae, until Fowler (1974) and Knaff and Buchanan (1975) identified them in $Chlorobium$ $spp.$ and $Chromatium$.

In $R. rubrum$, the cyclic system can be inhibited by antimycin A, which is a long-known inhibitor of cytochrome b in mitochondria. Cytochrome c_2 is extractable, also inhibiting the system; both inhibitions are relieved by addition of the dye phenazine methosulphate (PMS). These ideas are represented diagrammatically in Figure 8.7.

The relationship of electron transport to phosphorylation will be discussed in Chapter 9.

Bacterial photosynthesis is clearly similar in many ways to system I of green plants. The pigment complex has a low fluorescence, and a barely detectable

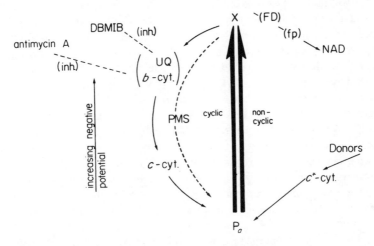

Figure 8.7. General scheme for bacterial photosynthetic electron transport. The split light-arrow indicates the possibility (only) of two separate centres driving the two established pathways. The b- . c^* and c-cytochromes are species-specific. The order of UQ and b-cyt is discussed in the text. FD, fp represent the ferredoxin–NADP–reductase pathway, not established in all types. The potentials of the P_a are of the order of 0·5 V; that of X is not indicated and is discussed in the text

delayed light emission at room temperature, and the active centre bacteriochlorophyll (P_a) not only resembles P700 but is also reduced by electrons from a c-type cytochrome. The involvement of the b-cytochrome makes the bacterial cyclic system similar to the cyclic scheme for system I in green plants (Knaff & Arnon, 1969b) (Figure 8.6), although the more important alternatives have less in common. Other important differences between the photochemical apparatus of plants and bacteria are the apparent lack of need for ferredoxin, at least in the cyclic system, the appearance of ubiquinone (the green-plant analogue, plastoquinone, being a system II material) and the lack of evidence for an analogue of plastocyanin between the c-cytochrome and P870.

CHAPTER 9

Phosphorylation

Phosphorylation, meaning the formation of ATP from ADP, is a ubiquitous and obligatory biological activity: it is the link by which catabolic metabolism is coupled to the endergonic processes for the maintenance of the cell. 'ATP is the energy currency of the cell' (R. Hill). In such processes as the aerobic respiration of glucose to carbon dioxide and water by the pathways of glycolysis and the citric acid cycle, 38 moles of ATP per mole of glucose are formed (two are consumed), and of these 38, 32 are formed by the oxidation of reduced coenzymes by oxygen in mitochondria. This is *oxidative phosphorylation*. The remaining 6 moles of ATP are formed by enzymes that react with metabolic substrates (for example, pyruvate kinase) and this is termed *substrate-level* phosphorylation. The two processes together are sometimes called oxidative phosphorylation, and the term *respiratory-chain-level* phosphorylation used to denote the mitochondrial part. The mechanism of substrate-level phosphorylation is relatively well understood, whereas there is no molecular model in sight that can explain the latter activity. When ATP was found to be a product of chromatophores and chloroplasts in light, from the researches of Frenkel and of Arnon, its connection with electron transport in a membrane established *photophosphorylation* as an analogue of respiratory-chain-level phosphorylation, and in this chapter we shall consider the two together. Certainly they are much alike, and there has already been considerable progress made by relating the ideas derived from the study of one into the study of the other.

9.1 Oxidative phosphorylation

Figure 4.4 sets out on a redox-potential diagram a sequence of carriers which probably represent the mitochondrial electron transport chain. Passage of two electrons (reducing one half-molecule of oxygen) from NADH to one half-molecule of oxygen results in the formation of three molecules of ATP. This 'P/O' or 'P/2e$^-$' ratio on the other hand is only two when succinate is the electron donor. This is taken to indicate that there are three phosphorylation sites on the NADH pathway, of which two are common to the succinate pathway. Further location of these sites is arrived at by three lines of reasoning. First, inspection of Figure 4.4 shows wide potential gaps between adjacent redox carriers in three or possibly four cases: NADH–FAD, FAD–ubiquinone, ubiquinone–cytochrome c_1 and cytochrome a–oxygen. Secondly, by applying reagents which inject or remove electrons from more or less specific sites in the chain, one can measure the P/2e$^-$ ratio in the part of the chain thus isolated. The results from such experiences are illustrated in Figure 9.1. Thirdly, the *crossover* method of Chance makes use of the inhibition of electron transport in mitichondria supplied with P_i but deficient in ADP (state 4). When ADP is added, electron transport accelerates and the steady-state redox level of the redox carriers is altered, as

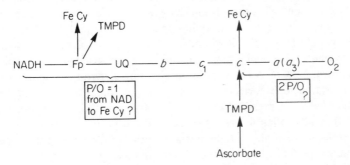

Figure 9.1. Diagram of the mitochondrial electron transport pathway, showing its interactions with artificial electron acceptors and donors, and the P/O ratios obtained. However, note that the value of 1 for the section NAD–FeCy is to be considered as an approximation to 2, and the value of 2 for the latter part is considered to be brought about by a contribution from the section NAD–Fp–TMPD, even in the presence of the inhibitor amytal. FeCy: ferricyanide; TMPD: tetramethylphenylenediamine. After M. Klingenberg (1968). In T. P. Singer (Ed.), *Biological Oxidation*, Interscience, New York, p. 3, with permission

observed spectroscopically. Redox carriers to the oxygen side of the substrate side become more oxidized. Thus the 'crossover point' indicates the rate-limiting phosphorylation site. The entire observation, repeated in the presence of low and high concentrations of an inhibitor such as azide shows the three sites indicated in Figure 9.2.

It should be noted that at least part of the electron transport pathway of mitochondria is reversible when coupled to phosphorylation. Given ATP,

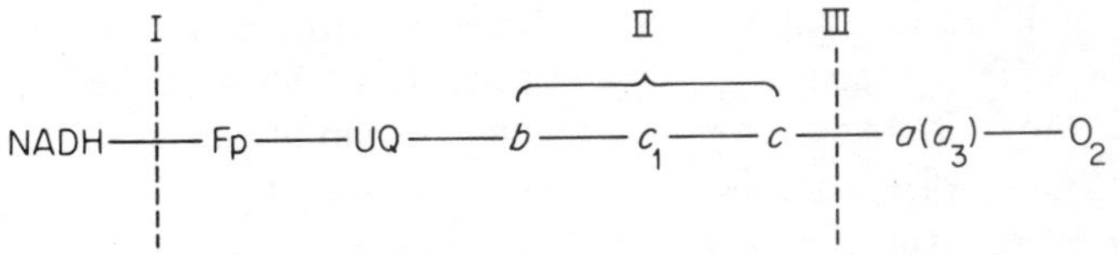

$$NADH \text{—} Fp \text{—} UQ \text{—} b \text{—} c_1 \text{—} c \text{—} a(a_3) \text{—} O_2$$

Figure 9.2. As in Fig. 9.1, showing phosphorylation sites deduced from the 'crossover theorem' (see text)

NAD^+ and succinate, mitochondria can form NADH (and fumarate), splitting the ATP which provides the energy necessary to drive the electrons to the more negative potential. The process is known as *reverse electron transport*. In the same way, an *energy-linked transhydrogenase* can be demonstrated which reduces $NADP^+$ using ATP and NADH. Both these activities can be driven by the energy released by the oxidation of succinate by oxygen in the remainder of the respiratory chain, the energy being transferred in some form other than ATP (e.g. $X \sim I$). Both processes are known in the chromatophores of photosynthetic bacteria, using the energy of light, ATP or pyrophosphate.

Evidence regarding the mechanism of ATP formation is derived from various kinds of experiment. First there is the study of *uncouplers*, which accelerate electron transport in state 4 mitochondria, and reduce the P/O ratio in phosphorylating (state 3) mitochondria. Typical uncouplers are arsenate and 2,4-dinitrophenol. Secondly, there are inhibitors such as oligomycin which prevent the formation of ATP without accelerating electron transport in state 4 mitochondria; in state 3, electron transport is slowed down, so that P/O ratio is not reduced. In the presence of both oligomycin and dinitrophenol only the effects of dinitrophenol are observed. Thirdly, *ion movements* can be demonstrated in respiring mitochondria, and these take place at the expense of the formation of ATP. The mitochondrion can take up approximately two Ca^{2+} ions (depending on the conditions) per half-molecule of oxygen reduced; $6H^+$ ions are extruded at the same time when the reductant is NADH, and $4H^+$ with succinate. Strontium (Sr^{2+}) and Mn^{2+} ions behave similarly. The uncoupler valinomycin permits the uptake of K^+ ions on the same scale. Oligomycin does not inhibit these ion-movements when they are driven by respiration but it does so when the same movements are driven by exogenous ATP in non-respiring mitochondria.

A fourth line of attack on this problem derives from the preparation of *coupling factors* from the small stalked spheres that can be seen on the cristae of mitochondria and removed by ultrasonic vibration with ethylenediamine tetraacetate. These stalked spheres were illustrated in Plate 9 (p. 73). From these particles proteins have been isolated and termed coupling factors since they restore the phosphorylating potency to inactive material (see Racker and Conover, 1963). The factor F_1, a protein of molecular weight 280 000, has Mg^{2+}-dependent ATPase activity. ATPases constitute a fifth line of attack in themselves, since their activity is likely to be a deranged form of an ATP-utilizing system, or the ATP-synthesizing system itself. Here it is noted that when the F_1 factor is combined with the F_0 factor (obtained from the mitochondrial membrane), the ATPase becomes sensitive to oligomycin. These stalked spheres are hence likely to be intimately connected with phosphorylation.

As remarked earlier, there are two principal types of theory which attempt to rationalize these results, the *chemical intermediate* and the *chemiosmotic* formulations. Both of these theories recognize that the energy released by the oxidoreduction reaction step must be conserved in at least two forms before either P_i or ADP is combined, and ATP formed. The difference lies in the postulated form that these energy stores take. The chemical intermediate theory looks for identifiable complexes of each of three redox carriers C_1, C_2 and C_3 with intermediates I_1, I_2 and I_3; these complexes CI are high energy complexes, and are often represented by a tilde ($\sim$) thus: $C \sim I$. The chemical intermediate theory continues by supposing that C is displaced by another unknown reactant,

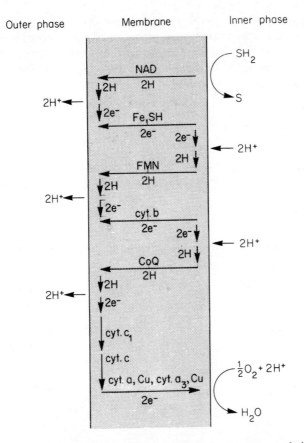

Figure 9.3. An arrangement of the mitochondrial electron transport chain, drawn to demonstrate means for the forced translocation of H^+ ions across the membrane. From P. Mitchell (1966). *Biol. Revs.* **41**, 445, with permission

X, giving a second high energy complex, $X \sim I$. I can then be displaced by P_i (or possibly ADP) giving $X \sim P$, and finally X is displaced by ADP giving ATP (see Figure 4.11). X and possibly I may be common to all three phosphorylation sites, or there may be site-specific intermediates X_1, X_2, X_3 and so on.

The *chemiosmotic theory* put forward by Mitchell (see Mitchell, 1966), regards the first energy store not as a chemical complex but as an electrochemical gradient across the membrane. In its simplest form this could be a pH gradient formed by pumping H^+ ions out of the mitochondrion during electron transport. The H^+ ions could then be exchanged with, say, K^+ ions so as to introduce an electric potential difference, the combined effect of the chemical and electric gradients being the 'protonmotive force'. Mitchell points out that this could be achieved by arranging the chain in the membrane according to Figure 9.3. Some of the redox carriers of the chain are of the A/AH_2 type: when these are reduced by a carrier of the A/A^- type a proton must be taken up from the aqueous phase, and similarly, when AH_2 is oxidized by an electron carrier, a proton is released. Mitchell suggested that these processes take place at opposite sides of the membrane so that protons are pumped across. The next energy store in this hypothesis is a complex $X \sim I$ which shuttles across the membrane reacting on one side with ADP and P_i giving ATP:

$$X \sim I + ADP + P_i \xrightarrow{\text{alkaline pH}} X^- + IO^- + ATP + 2H^+$$

$$\text{or } XH + IO^- + ATP + H^+$$

$X \sim I$ has the properties of an anhydride (or possibly an ester). At the other side of the membrane X^- and IO^- condense to reform $X \sim I$ taking up protons:

$$X^- + IO^- + 2H^+ \xrightarrow{\text{acid pH}} X \sim I + H_2O$$

Strictly, XI is only a high-energy compound in the alkaline environment. The two reactions together constitute either an ATP-driven H^+ ion pump, acting in

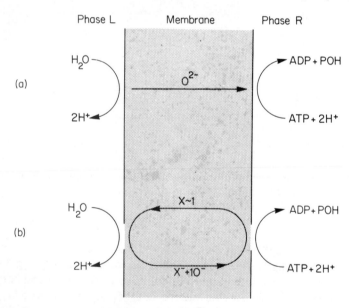

Figure 9.4. Proton-translocating reversible ATPase system. A: principle; B: operation with a hypothetical carrier, the anhydride $X \sim I$. From P. Mitchell (1966). *Biol. Revs.* **41,** 445, with permission

reverse, or an H^+-driven ATP synthetase, shown in Figure 9.4. The magnitude of the gradient necessary is at least 0·21 V (electric potential) or 3·5 pH units (chemical driving force), or a combination of the two, depending on the conditions; for a two-electron, or two-proton, transfer, this value converts to 41 kJ mole^{-1} (9·7 kcal mole^{-1}), which is an approximate minimum for the formation of ATP under physiological conditions.

The conformational theory of phosphorylation, put forward by Boyer (1965), suggests that electron transport, at the coupling sites, leaves certain protein molecules in a conformationally-stressed condition. This stress (the first store of energy) is transferred (by stressing intermediate proteins) until it reaches and stresses the F_1 particle. The F_1 particle, whose groups might be identifiable with X, I, etc., relieves its stress in forming ATP. This hypothesis has been enthusiastically taken up by Green (1974) who has extended it into the 'electromechanical' hypothesis. This makes use of some of the key features of Mitchell's ideas, principally the idea of vectorial electron transport across a membrane and alternation of hydrogen and electron carriers. The operation of each unit of the mitochondrial electron-transport chain (which can be extracted in four complexes) produces electric fields which set up the stress. The F_1 coupling factor migrates so as to pick up conformational energy.

Although in some circumstances the conformational hypothesis can be regarded as a special case of the chemical intermediate hypothesis, for example if $C \sim I$ were an internal bond formed within a single protein, there is an important difference. The energy of formation of a ($\sim$) covalent bond would be expected to have a fixed value under set conditions (there cannot be a fractional bond) and therefore the energy taken from the redox reaction is quantized. Conformational energy is less clearly defined—there may be two or more discrete conformations with fixed energies, or the system may resemble a continuously-compressible spring. Energy transmission via hypothetical matched series of discrete conformational state changes has been termed *conformon* transmission. Apart from the observations of new spectral peaks (below), the conformational hypothesis has not so far suggested any discriminatory experiment. It has however been claimed that the contractile proteins obtained from mitochondria and chloroplasts operate in this role.

It is essential that there should be a 'sidedness' in phosphorylating membranes if the chemiosmotic hypothesis is correct. In the mitochondrion, ion-movements suggest that if the chemiosmotic theory holds then the inside of the mitochondrion is rendered alkaline by electron transport, hydrogen ions being expelled. Work done on vesicles prepared from mitochondria with digitonin appears to contradict this, but electron microscopy shows that the stalked sphere ('coupling factors') are now on the outside of the vesicles, in other words the membrane is inside-out. This must be borne in mind when discussing the chloroplast.

9.2 Photosynthetic phosphorylation

9.21 Phosphorylation sites

This concept was inherited from the time when mitochondrial phosphorylation coupled to electron transport was the only form known. There were good grounds for associating each molecule of ATP produced with a specific section of the respiratory chain. This is in its extreme form in the chemical-intermediate hypothesis. The chemiosmotic hypothesis however requires sites for energy conservation, such that electron flow through a site (which may cover a greater or lesser section of the electron transport pathway) causes the passage of hydrogen ions against a pH gradient. The conformational scheme originally involved the idea of certain redox carriers (at specific sites) acquiring conformational energy from the redox action, and support was found for this in the change of redox potential of cytochromes $a–a_3$ observed with ATP, and a similar change observed with components of cytochrome b (see Wilson and coworkers, 1973; Erecinska and coworkers, 1973). However Green (1974) in his review of the electromechanical extension of the conformational idea prefers to regard all electron-carrying proteins as capable of acquiring conformational energy since in either the oxidized or reduced condition the unbalanced charge must generate stress. This is taking a somewhat free line, however, since some electron-transfers in the chain are susceptible to the back-pressure developed when phosphorylation is inhibited by lack of ADP. This was established in the chloroplast for the reduction of cytochrome f by Avron and Chance (1966) using essentially the crossover technique. If the zig-zag scheme of non-cyclic electron transport is sketched on a scale of standard redox potentials, it is seen that the step from plastoquinone to f represents a jump of some 0.3 V (there is uncertainty in the potential of plastoquinone), much bigger than any known jump elsewhere. This jump of 0.3 V is a reasonable value for a step which generates ATP, since it would allow some 59 kJ (14 kcal) of energy to be extracted (for the passage of two electrons).

The variable redox potential of cytochrome b-559 attracts attention by analogy with the mitochondrial b_T, but, as described in the previous chapter, kinetic studies have so far failed to find a role for it.

Other potential jumps may be concealed in the inaccessible regions of electron transport, however; the potential of the site Y_{II} is not known at all and might allow a jump of sufficient magnitude between Y_{II} and oxygen-evolving system. The same applies to the acceptor X_{II}, and the step X_{II} to plastoquinone. System I is better known and offers little possibility. To resolve these doubts there have been studies of the number of ATP molecules formed in non-cyclic electron transport (the $P/2e^-$ ratio), the number of H^+ ions transferred (H^+/e^- ratio) and the number of H^+ ions required to synthesize an ATP molecule (H^+/P ratio).

Stoichiometry of phosphorylation. Direct measurements are subject to the problem that chloroplasts with their outer envelopes intact are virtually impermeable to adenine nucleotides. (The next chapter discusses the effects of the envelope on metabolism, etc.) Using envelope-free chloroplasts measurements have been made of the $P/2e^-$ ratio in non-cyclic systems evolving oxygen and reducing one of the following: NADP (in the presence of added ferredoxin) ferricyanide, oxidized DCPIP, diaminodurene (DAD, 2,3,5,6-tetramethyl-*p*-phenylenediamine) and others. Values in excess of 1 have been widely recorded. On the old concept, the number of coupling sites had to be a whole number, so the number of molecules of ATP formed was therefore considered to be 2, with experimental losses. However, the suggestion could not be avoided that these losses were natural, that is, they occurred in the natural chloroplast, or alternatively that, according to the chemiosmotic hypothesis, the independence of the $H^+/2e^-$ and the P/H^+ ratios made fractional $P/2e^-$ ratios quite possible. It is apparent from Chapter 5 that the reductive pentose cycle cannot work as described if there are less than $1 \cdot 5$ ATP molecules per NADPH (that is a $P/2e^-$ ratio of $1 \cdot 5$), and a slightly higher figure is needed to allow for ATP consuming processes such as the formation of starch, sucrose, etc. Any deficit, however, was expected to be filled by a contribution from cyclic photophosphorylation.

The $P/2e^-$ value of approximately $1 \cdot 4 - 1 \cdot 7$ reported by Reeves and coworkers (1972) could be raised to $2 \cdot 0$ by subtracting the basal rate of electron transport in the absence of ATP production from that which took place during phosphorylation (an interpretation made and defended by Izawa and Good (1968). This was argued (Reeves and Hall, 1973) to indicate two coupling sites when ferricyanide was the acceptor; when DAD was used the figure was approximately half. Ferricyanide accepts electrons from system I in relatively undamaged preparations, since being water-soluble it has no access to the chain between systems I and II. DAD is lipid soluble and accepts from system II. Hence the above authors claimed one coupling site in system I and the other in system II. The two acceptors are examples of classes I and III respectively in the scheme of Saha and coworkers (1971).

An indirect approach was made by Heber and Kirk (1974), who used intact chloroplasts, provided with substrates that required different proportions of ATP and NADPH for their reduction. They found the quantum requirement for the reduction (consuming 1 molecule of NADPH in each case) was four for oxaloacetate (no ATP required), four for phosphoglycerate (1 ATP) and six for glycerate (2 ATP); 9–11 quanta were required for glycerate reduction at higher pH values. They concluded that coupling in non-cyclic electron transport was variable according to the demand for ATP, up to a maximum $P/2e^-$ ratio of $1 \cdot 1 - 1 \cdot 4$. When extra ATP was needed this had to be provided from a different source, using more quanta. They found no evidence for a basal uncoupled rate of electron flow.

There is of course good ground for believing in cyclic photophosphorylation; for example, Tanner and Kandler (1969) reported that *Chlorella* cells could take up glucose. This required energy, calculated to be 2ATP/glucose, which could be provided by light of 712 nm wavelength (system I, hence cyclic). The quantum requirement was $4 \cdot 3$ quanta per glucose, suggesting a $P/2e^-$ ratio of 2 for system I in the cyclic mode. The need for only one coupling site in cyclic electron transport presents no problems with any of the schemes offered.

Coupling sites in partial reactions. Although the reaction between plastoquinone and cytochrome f is indicated as a coupling site both on kinetic grounds and because of its potential jump, the concept became confused when the coupling of the partial reactions of systems I and II were studied. Thus the donor system ascorbate-DCPIP provides electrons to system I, probably via plastocyanin, and electron transport can be measured with either ferredoxin and NADP, or methyl viologen. In each case the rate is markedly increased by the addition of un-couplers, and ATP has been shown to be produced when either acceptor was used. There is no obvious classical coupling site for phosphorylation in this pathway. In the same way a partial reaction around system II can be constructed using phenylenediamine (or C-substituted phenylenediamines such as DAD) which is lipophilic and accepts electrons from plastoquinone. Such electron acceptors are termed class III by Saha and coworkers (1971); usually they are added in catalytic quantities and are kept oxidized by ferricyanide. Water is the electron donor to system II. ATP is produced, with a $P/2e^-$ ratio in the range $0\cdot4$–$0\cdot7$, i.e. half that of the complete electron transport pathway. Once again the obvious coupling site, PQ–cyt.f, is not used, and no other potential jump is easy to find in the electron-transport pathway around system II.

9.22 The chemiosmotic account

Trebst (1974) has reviewed the evidence connecting electron transport with hydrogen-ion uptake by thylakoids, and drawn the scheme (Figure 9.5) in which

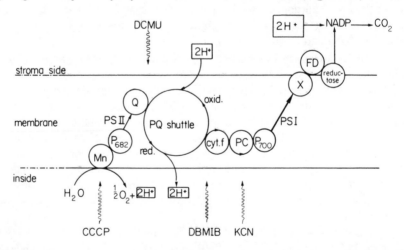

Figure 9.5. Arrangement of the non-cyclic electron-transport scheme in order to explain coupling in terms of H^+ translocation from the stroma to the inside of the thylakoid sac. The wavy arrows indicate the point of action of four inhibitors. From Trebst (1974). Reproduced, with permission, from *Annual Review of Plant Physiology*, Volume 25. Copyright © 1974 by Annual Reviews Inc. All rights reserved

a consensus version of the non-cyclic process (Figure 8.2) is arranged so as to explain (i) the accessibility of sections of the chain to reagents acting on one or other side of the thylakoid membrane, (ii) the flow of hydrogen-ions into the sac with stoichiometry such that $H^+/2e^- = 4$, and (iii) the directional nature of the primary process ($P_aX \rightarrow P_a^+ X^-$) such that charges tend to be moved *across* the

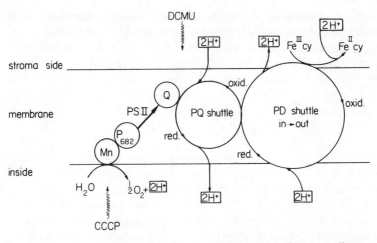

Figure 9.6. Functional isolation of system II, preserving coupling, using phenylenediamine (PD) and ferricyanide (Fe^{III}cy). Note that $2 \times 2H^+$ are translocated, but $2H^+$ are lost. From Trebst (1974). Reproduced, with permission, from *Annual Review of Plant Physiology*, Volume 25. Copyright © 1974 by Annual Reviews Inc. All rights reserved

membrane. This scheme offers solutions for the puzzles raised by the coupling of the partial reactions described in the last section, shown in Figures 9.6 and 9.7. The key factor is the lipid solubility of the reagents, allowing them to shuttle backwards and forwards across the membrane, thus forming part of both the electron-transport and hydrogen-ion translocating systems.

The partial-system I reaction scheme 9.6 is clearly adaptable to explain cyclic electron transport around system I, and its coupling. When artificial agents such

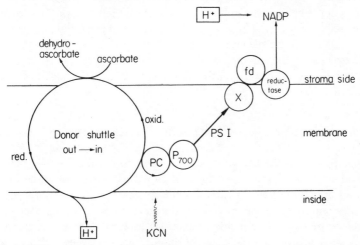

Figure 9.7. Functional isolation of system I, preserving coupling, using ascorbate and a donor such as DCPIP. From Trebst (1974). Reproduced, with permission, from *Annual Review of Plant Physiology*, Volume 25. Copyright © 1974 by Annual Reviews Inc. All rights reserved

as PMS are used, they act as electron and hydrogen-transporting shuttles (Figure 9.8); it makes no difference from this point of view whether they react with cytochrome f, plastocyanin or P700, since these are all electron carriers. If ferredoxin is used as a 'natural' cyclic phosphorylation cofactor, being lipophobic (a class I acceptor) it cannot form a shuttle, and so Trebst suggests that it acts via plastoquinone. The two cases can be distinguished as shown by sensitivity to the inhibitor of plastoquinone oxidation, DBMIB (Hauska and coworkers, 1974). In the partial systems shown, and in the cyclic systems, $P/2e^- = 2$.

The conversion of hydrogen-ion movement to ATP is not explained. At the present time it can be described in phenomenological terms. Witt (1971) reviewed work in which the H^+/P ratio was evaluated as $3 \cdot 2$, taken as 3. Other groups have claimed 2, 3 and 4, the later reports tending to be the higher. Graeber and Witt (1974) found values which depended on the magnitude of the protonmotive force (combined pH and membrane potential); the lowest values were not less than 3. There is no answer to the question of how many H^+ are required by the actual ATP synthetase mechanism, and how many leak past, either through natural imperfections or experimental damage. As we shall see later, the actual synthetase from chloroplasts does not appear to interact with hydrogen-ion gradients, so that another catalyst will have to be elucidated to connect the two. Until the mechanism of this connection is established in molecular terms, it will not be possible to answer the questions raised in the section on the stoichiometry of photophosphorylation. Meanwhile there does appear to be ground for believing in a fractional coupling ratio, $ATP/2e^- = 1 \cdot 33$ at best. The contribution from a cyclic or pseudocyclic system therefore retains its theoretical importance.

Shahak and coworkers (1975) have reported an oxidation of cytochrome f caused by a pH change in a chloroplast suspension in the dark. This appears to be an example of *reverse electron flow* between f and plastoquinone, the energy for which comes from a pH gradient of $3 \cdot 8$ units. This is a further demonstration, if one were needed, that such gradients can be coupled to redox processes (and to ATP), but it does not prove that the chemiosmotic form is obligatory in energy transmission.

Generation of electric potential. One of the features of Figure 9.5 was that the primary processes were oriented so that the electron movement from the photoreactive P_a to the acceptor X induced a potential difference across the thylakoid membrane. That this field is developed comes from study of a spectral change in the region 515–518 nm in chloroplasts or 520–530 nm in bacteria. This absorbance change can be produced by a single flash, which operates each reaction centre once only. Under conditions that only allow the operation of either system I or system II, the absorbance change is only half its previous value. The speed of development of this effect is faster than can be measured (less than 20 ns), which is also the case for the appearance of P_a^+. It should be borne in mind that flashes brighter than a certain threshold value produce a similar

169

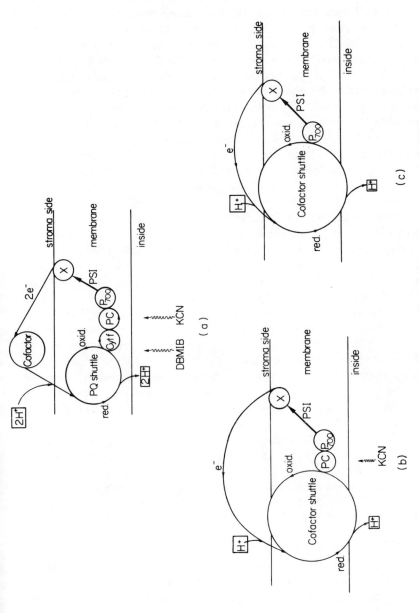

Figure 9.8. Cyclic electron transport. (a) The natural process (cofactor = ferredoxin) and, (b) and (c), artificial systems (possible cofactors DCPIP or DAD in (b), PMS in (c)). From Trebst (1974). Reproduced, with permission, from *Annual Review of Plant Physiology*, Volume 25. Copyright © 1974 by Annual Reviews Inc. All rights reserved

change in the carotenoid pigments that Witt terms a 'valve reaction', where excess energy is removed from the chlorophyll mass and deactivated.

Witt's group found that the 515 nm change (ΔA) decayed during the dark interval between flashes, and that the rate of the decay could be accelerated by (i) damage to the membranes (ii) hydrogen-carrying uncouplers and (iii) ionophores such as gramicidin D which make the membrane permeable to alkali–metal cations. Jackson and Crofts (1969), studying the corresponding changes in bacteria, found that ΔA could be produced by artificial membrane potentials, generated by placing the cells in solutions of potassium ions of various concentrations, in the presence of valinomycin, an ionophore of group (iii) above. Problem 5 (p. 93) was constructed by Dr. A. R. Crofts from such data. The argument is clear that the ΔA is due to a membrane potential (and directly proportional to it) and both Jackson and Crofts, and Witt (1971), presented a scale of relation. Such a potential across the small thickness of the thylakoid would amount to a field of some 10^5 V/cm. In such fields, chromophores show changes in their absorption spectra known as electrochromic shifts, or the Stark effect. The ΔA observed in chloroplasts corresponds to changes observed in multilayers of chloroplast pigments prepared on glass slides and subjected to such potential differences. (There is no special significance in the wavelength 515 nm apart from instrumental convenience.) Witt's group also calibrated the electrochromic response of the lipid-soluble dye rhodamine B and, using it as a vital stain, found the same change in illuminated chloroplasts.

The potential calculated from ΔA in one experiment was 50 mV when one flash was applied and both systems were active, and 25 mV when one of the two systems was isolated by the experimental conditions. A rapid succession of n flashes which reduced most of the plastoquinone pool caused an n-fold greater ΔA in the contribution from system II (n from 1 to 7). Transient potentials of 200 mV have been observed, with typical steady states in the region of 100 mV.

The field is collapsed by the action of gramicidin and potassium ions. The minimum concentration required for this depends on the quantity of thylakoid material present. One gramicidin molecule is active against 10^5 molecules of chlorophyll, which correspond to an area of 50 nm^2, the size of an average thylakoid. The field is therefore a property of the whole thylakoid sac. The same concentration of gramicidin inactivates phosphorylation, implying a connection between the electric field and ATP synthesis. Baltscheffsky and Hall (1974) showed that ΔA was diminished by ADP under phosphorylating conditions, supporting the hypothesis.

As shown in Figure 9.5, the translocation of hydrogen-ions by the electron-transport chain is *electrogenic*: not only a pH gradient but also a membrane potential would be developed. This provides a further source of membrane potential, which replaces that generated by the primary process (since the passage of electrons from plastoquinone to P700 is across the membrane the other way, thus reducing the field). A relative adjustment of the pH gradient and the field can be achieved by *field-driven H$^+$-expulsion* (see Witt, 1971), and also by the slower process of *field driven K$^+$-expulsion*. Under some conditions Witt's group found a correlation in the rate of decay of the field, and the rate of development of the pH gradient (as measured by the fluorescence of UBF). The hypothesis arrived at is that the primary processes generate a field, which is increased by the electron-transport process concomitantly with the production of a pH gradient. Energy in the field can be used to increase the pH gradient. Energy in the field can be stored by expulsion of metal ions from the thylakoid. (Such ion

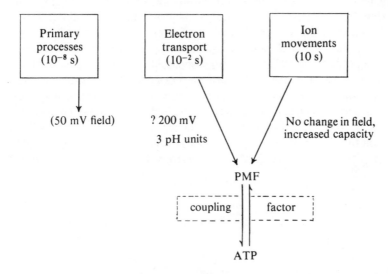

Figure 9.9. Hypothetical relationships of electron transport, ion movements and the formulation of ATP by means of an electrochemical energy store (protonmotive force: PMF)

movements will be discussed in section 9.4). Both field and pH gradients can be coupled to ATP synthesis (Figure 9.9.)

Natural and artificial pigmented membranes. In addition to pigmented membranes in plants and photosynthetic (anaerobic) bacteria, a pigmented membrane has been found in aerobic bacterium from a saltmarsh habitat, *Halobacterium halobium*, which in common with many halophytic bacteria, accumulates carotenes. In this case the purple colour is *bacteriorhodopsin*. In light, hydrogen-ion movement takes place which appears to be convertible into ATP formation. The bacterium therefore relieves its dependence on aerobic metabolism, at least to a large extent (Danon and Stoeckenius, 1974). The phenomenon has an interesting bearing on the concept of photosynthesis!

Secondly, it is possible to form a bi-layer lipid membrane across a small hole so as to separate two aqueous solutions. The thickness of such membranes is of the order of 5–10 nm. Pigments and dyes can be incorporated in these membranes and, when they are illuminated, electrons can be transferred from a reductant in solution on one side, to an oxidant in solution on the other, resulting in the generation of a potential difference between the two sides (Tien, 1974).

These observations make the generation of an electric field in the absence of an electron-transport system seem reasonable in photosynthesis.

9.23 The relative importance of the chemiosmotic and other hypotheses

It is widely accepted that chemiosmotic energy can be formed at some stage between the electron-transport reactions and the formation of ATP, and that this is reversible. This is established by, first, the experiment of Hind and Jagendorf (1963) that changing the pH of the suspending medium allowed ATP formation, leading to the concept of 'acid-base phosphorylation' (this work was crucial in extracting Mitchell's (1961) paper from apparent oblivion). Secondly the role of uncouplers is easily explained by their power of dissipating the pH gradient (by making protons soluble in the membrane) or neutralizing it (as with ammonium ions). Thirdly, two lines of work show a drop in the intra-thylakoid pH during illumination: these are the kinetic experiments of Rumberg and Siggel (1969) based on the sensitivity of the plastoquinone–cytochrome f reaction to pH, and the use of the fluorescent indicator umbelliferone (UBF) (see the reviews of Witt (1971 and 1975)).

non-fluorescing   Umbelliferone   fluorescing
(UBF)

Fourthly, the formation of ATP is correlated with the disappearance of the pH gradient (Rumberg and Siggel, 1969).

These findings do not of course establish the chemiosmotic hypothesis, which is that the combined pH gradient and membrane potential is an obligatory intermediate. The item of evidence having most weight in this respect is the fast rise time of the 515 nm ΔA, which is too fast to allow a credible explanation in terms of the prior formation of a covalent bond ($\sim$). The argument that a chemiosmotic energy store can be filled without any possibility of a prior ($\sim$), and then to form ATP, would establish the hypothesis. At present it is hardly complete. For example, although the primary-process potential arises so fast as to make a prior ($\sim$) unlikely, this potential plays no part in steady state phosphorylation; the important potential is that due to the redox-reaction hydrogen-ion translocation, which cannot be said to precede any possible ($\sim$).

Criticism of the chemiosmotic account. Cole and Aleem (1973) claim that a preparation from bacteria that has been reduced to a size too small to allow the possibility of vesicle formation can perform oxidative phosphorylation efficiently. Since a vesicle is essential for the chemiosmotic theory, this report strikes at the heart of it. However, it is a puzzling report since one would expect (X $\sim$ I) to be capable of operating the ion-movements that give rise to pH gradients and membrane potentials, and in the absence of a membrane this energy would be dissipated and the system should appear uncoupled. Either in this case there is no mechanism for operating ion-movements, or there are indeed vesicles, undiscovered.

The redox-potential changes in cytochromes a and b, originally taken as evidence of chemical intermediate formation ($C \sim I$) or conformational changes, are complex in nature, and can be accepted by any of the hypotheses (see the review by Dawson and Selwyn, 1974).

There is much less data available concerning the mitochondrion in respect of the location of electron-transport intermediates on one or other side of the cristal membrane. Whereas Trebst (1974) listed many observations for the chloroplast, the only observation in mitochondria appears to be the greater effectiveness of azide (N_3^-) in inhibiting cytochrome oxidase on the inside of the mitochondrion.

The coupling factor particles (ATPase) can be prepared in a stable 'energized' state (Forti and coworkers (1972), see below). This energized state cannot be a pH gradient or an electric field since there is no membrane, still less any vesicle. Therefore it must be either a high-energy covalent linkage ($\sim$) or a conformational state, or probably both. There is no data concerning the factor which energizes the ATPase particle, nor the source of the energy. The reader is referred to a recent review of Jagendorf (1975) for a comprehensive study of the mechanism of photophosphorylation.

Conclusion. A scheme of photosynthesis has been presented and worked out in some detail for conservation of electron-transport energy in a chemiosmotic form. There is no compelling evidence that the chemiosmotic form is an obligatory step in the eventual synthesis of ATP *apart* from the fast kinetics of the field formation. The last step in ATP synthesis is probably based on chemical (($\sim$) or conformation) energy; schemes based on the transduction of energy from electron transport entirely in the form of chemical energy (without the need for a chemiosmotic step) are available but lack the detail and the general appeal of the former. In systems other than the chloroplast, the chemiosmotic account is weaker.

9.3 The formation of ATP

ATPases and exchange reactions. Since ATP is such an important material, ATPases, which merely hydrolyse ATP to ADP and P_i, are unlikely to be of biological value to the cell, and therefore are probably artifacts produced by the experimental conditions, representing enzymes which had previously been concerned with ATP utilization or synthesis. Thus the ATPases of nervous tissue in animals may be derived from the ATP-requiring sodium pump; and in section 9.1 the coupling factors of Racker which have ATPase activity have been considered as part of the ATP synthesizing machinery of both mitochondria, chloroplasts and bacteria.

Unlike mitochondria, intact thylakoids do not show any ATPase activity in the dark. However a Ca^{2+}-dependent ATPase can be obtained from thylakoids;

this enzyme is inhibited by the antibiotic Dio-9 (which blocks ATP formation in the chloroplast). This ATPase appears to be derived from the coupling factor. A second ATPase was observed by Avron, in chloroplast material, which required both calcium and (continuous) light. A third chloroplast ATPase requires magnesium ions, and needs to be induced by either a brief illumination or an acid–alkali transition in the presence of reducing agents such as dithiothreitol. This is termed the light-triggered ATPase. The effect of the triggering dies away rapidly unless ATP is present, when the ATPase activity persists unabated for many minutes.

This can be explained by supposing that the formation of ATP catalyzed by the coupling factor is reversible, and that there is an equilibrium between ATP and the energy store (which we believe to be the PMF across the thylakoid membrane). If the system were detached from the thylakoid, the energy would be dissipated and an ATPase action would result. This provides a rationalization of the ATPase of the coupling factor. Secondly, in the intact thylakoid, uncoupling agents which are believed to render the membrane permeable to H^+ ions and so collapse the PMF would bring about the hydrolysis of ATP in the same way. This is in fact observed in mitochondria, but not in chloroplasts. However, uncoupling agents do have two effects in the chloroplast in this context: first they inhibit the triggering of the light-triggered ATPase, and secondly they stimulate it once it has begun. We could conclude from these results, and from the need for ATP to maintain the light-triggered ATPase, that ATP has a controlling action on the phosphorylating enzymes. 'Allosteric' effects of ATP on other enzymes are well known. A biological reason for such an effect is that plants must face regular periods of darkness, when they obtain their ATP from mitochondria. This 'switching off' of the chloroplast would save wasting ATP. In mitochondria activity may be continuous so that external ATP is always balanced by 'respiration pressure'. It does not therefore need to have such a switching-off mechanism.

Further information can be obtained from observing *exchange reactions*. This type of reaction is carried out under conditions where, although no net changes take place in the quantity of the materials present, isotopic label is transferred from one substance to another. Thus if thylakoids, ATP and radioactive inorganic phosphate ($^{32}P_i$) are incubated together, some of the radioactivity may become incorporated into the ATP although the total quantity of ATP hardly changes. This is the ATP–P_i exchange, and it is significant that it needs to be triggered in the same way as the light-triggered ATPase. An ATP–P_i exchange would be expected from the formation of $X \sim I$, with the splitting of ATP to ADP and P_i. The P_i would mix with the radioactive P_i and the reverse reaction would re-synthesize ATP containing some ^{32}P. This locates the control site at that particular reaction.

Models proposed. The general model of Boyer (1968) (Figure 9.10) provides a 'hat-stand' on which to arrange the phenomena reported. The box indicates two states of a particle, ATP synthetase or ATPase, which are interconvertible given any energy input or drain. There is no restriction on the form of the energy source. All the reported exchange reactions are accounted for. The only problem is that the P_i–H_2O and ATP–H_2O are much faster than the ATP–P_i exchange (Mitchell and coworkers, 1967). They suggested that either the P_i–H_2O exchange was a separate process, or that there was a second entry point for water.

Conformational changes have been strongly suggested to occur in the synthetase particle by Ryrie and Jagendorf (1971), since they found that exchange can take place between tritiated water and hydrogen atoms in the protein, but only when the particle was energized by ATP synthesis or hydrolysis. This indicated that these exchange sites were inside the tertiary structure of the protein

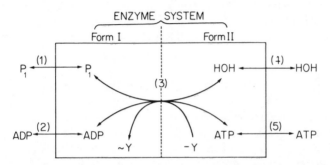

Figure 9.10. Scheme from Boyer (1968) showing a hypothetical enzyme interconverting two forms of a complex (E.P_i . ADP ⇌ E.HOH.ATP). Steps (1), (2), (4), (5) are the binding/dissociation reactions of P_i, ADP, H_2O and ATP respectively. Step (3) shows phosphorylation, driven by the energy reservoir ~Y, and the reverse ATPase reaction. Reproduced with permission

and were exposed during action—a clear case of conformational change. Again, Forti and coworkers (1972) found that fresh preparations of the coupling factor would incorporate ADP and P_i forming a new pyrophosphate bond. The protein must have carried the energy in a covalent (~) or conformational form (or both).

Polarity. It is interesting from the standpoint of the chemiosmotic hypothesis that photophosphorylation seems to be of opposite polarity to oxidative phosphorylation. Thus the chloroplast and chromatophore both accumulate hydrogen ions, whereas the mitochondrion extrudes them; thus the inside of thylakoids becomes more acid than the outside, whereas mitochondria, at least in the presence of calcium ions, become alkaline inside. The weak-base uncouplers of the chloroplast, the primary amines and ammonium salts, have little effect on the mitochondrion, which is the more powerfully affected by the weak-acid types. Lastly, the 9 nm stalked spheres (Plate 9, p. 73) occur on the outside of the photosynthetic membranes, but on the inside of the mitochondria.

Coupling factors. The 9 nm stalked spheres (see Plate 9, p. 73) appear to contain as a principal component the factor F_1 (in mitochondria) or CF_1 (in-chloroplasts)* (see McCarty and Racker, 1967). A list of protein factors related to oxidative phosphorylation in mammalian mitochondria is given by Lardy and Ferguson (1969). F_1 is an ATPase, and CF_1 becomes an ATPase after a brief activation by trypsin. Several materials of this type, listed by Lardy and Ferguson as phosphoryl transferases, have been isolated from mitochondria, and may well represent the final enzyme in the pathway that produces ATP in respiration. By analogy, CF_1 may be presumed to be the final enzyme involved in photophosphorylation. Although CF_1 has no ATPase activity before treatment with trypsin, neither does the chloroplast; ATPase activity can also be induced in chloroplasts by treatment with light or an acid-alkali transition (both of which are assumed to generate a high-energy state ($\sim$) in the presence of reducing agents such as dithiothreitol, and CF_1 similarly develops ATPase activity given dithiothreitol and thylakoid material. Both F_1 and CF_1 restore phosphorylation to membrane material from which the stalked spheres have been removed.

Conclusion. There is enough data to eliminate some of the enzymic models for the synthesis of ATP, but apart from that this process is as obscure as the photosynthetic production of oxygen (or the genetic code two decades ago). The isolation, crystallization and total crystallographic solution of the ATPase, as well as of the controlling protein that couples it to the chemiosmotic energy store, would provide a suitable task for heroes, and might finally settle the nature of the chemical energy($\sim$) form, the extent of conformational changes, the 'ideal' value of the ratio H^+/P and the exact molecular mechanism for the synthesis of ATP.

9.4 Long-term, energy dependent, ion fluxes

In this section we gather together a diverse collection of phenomena that can be related to the energy state (charge, level) of thylakoid membranes.

Lavorel (1959) in his classic study of the Kautsky effect recorded two stages in the development of *fluorescence*: an increase, 'I–P' usually complete in a few seconds and ascribed to the reduction of the quencher Q by system II (see Chapter 8), and a much slower fall 'P–S' that might take a minute. Murata (1969) showed that the S level could be increased by the addition of 2 mM magnesium ions (or 100 mM sodium). Sinclair (1972) found that approximately the same addition increased the *Emerson enhancement* in the water–NADP electron-transport system. Both workers concluded that these concentrations of ions diminished the ability of the chloroplast to transfer excitation energy from the chlorophyll mass of system II to that of system I.

Barber and coworkers (1974), using the ionophorous antibiotic A23187 which is specific for divalent cations, were able to show that magnesium ions were expelled over a period of several minutes from the thylakoids into the

* It so happens that *chloroplast* factors have been named CF_1, CF_2, CF_6, etc., and that CF_1 is analogous to F_1 in mitochondria. Kagawa and Racker have however allotted the name CF_0 to a preparation derived from the mitochondrial factor F_0 by means of cholic acid: to that extent the nomenclature is inconsistent.

stroma of illuminated chloroplasts. Previously Mayne (1969) had shown that *delayed light emission* was regulated by the high energy state of thylakoids, although over times of a few seconds. Other such studies (e.g. Bertsch and coworkers, 1969) stressed the time (more than 1 minute) required to reach steady conditions before recording the waveform. Gregory (1975) found a diminution in the chlorophyll–chlorophyll interaction observed with *circular dichroism* in whole chloroplasts, brought about over 1–2 minutes of illumination, which was interpreted as indicating an expansion of the thylakoid–thylakoid distances in grana.

This pattern of ionic effects can be related to the later stages in the formation of the chemiosmotic gradient, in which the electric field expels metal ions from the thylakoid (Figure 9.9) (Witt, 1971). (In Witt's original scheme, K^+ was used to stand for all ions other than H^+ or OH^-.) This field-driven efflux provides a reservoir increasing the capacity of the field while limiting its amplitude. The pH gradient is not affected. The energy of the efflux should be regained during phosphorylation if the efflux of hydrogen ions is coupled to an influx of the metal ions. This concept is equally valid whether the chemiosmotic energy store is formed as a direct or indirect step in phosphorylation.

The effects on the pigment system might be considered to be fortuitous changes, only serving an experimental convenience as indicators of ion fluxes, or they may in themselves have a regulatory role. This would operate perhaps in such a way that in dim light, or when cyclic photophosphorylation was in demand, the failing chemiosmotic field would, by resorbing the magnesium ions, alter the arrangement of thylakoid membranes and possibly of units within each membrane, so that quanta could migrate further and avoid being wasted on blocked reaction centres, and so that energy of excitation could be more easily passed from system II to system I (where it would operate cyclic phosphorylation). This rearrangement of pigment under 'chemiosmotic pressure' has been termed 'plastic chlorophyll'.

Lastly, a number of the enzymes of the reductive pentose cycle have been shown to be controlled by light, by pH and by magnesium ions. The property of thylakoids in extruding the latter under the influence of the former provides a tidy regulatory system.

CHAPTER 10

Chloroplast metabolism and its relation to that of the cell

10.1 The chloroplast envelope

So far in this text it has been possible to treat the chloroplast as a completely independent organelle. In this chapter topics which depend on an interaction between chloroplast and cytoplasm are gathered together. Such interactions are necessarily mediated by the passage of chemical materials across the envelope of the chloroplast. Plates 1 and 2 (pp. 12 and 13) show the double envelope clearly. There are reports that the inner membrane exercises most of the control over the passage of solutes, and that the outer layer is permeable to most small molecules. Studies with cell membranes in general, in particular with external cell membranes of cells such as the alga *Chara*, or erythrocytes, suggests that substances which have an appreciable solubility in lipid can pass across a membrane by a simple process of dissolving into the lipid component from one side and out on the other. Pores in the protein network are quite large and size is less important than the lipid solubility of the permeant. None of the metabolic substrates we have to consider in this section have an appreciable lipid solubility: many are ionized, for example Mg^{2+}, malate and glyceraldehyde-3-phosphate. It is usually the case that to pass across membranes such compounds need a translocator or permease system, which may be thought of as an enzyme-like .entity in the thickness of the membrane that binds the substrate on one side and carries it across to be released on the other. In the case of, for example, the red blood cell, it is relatively easy to show that the entry of radioactive sorbose increases in velocity with increasing concentration of sorbose in the medium, up to an eventual maximum rate, when the sorbose concentration is often said to be saturating. A Michaelis constant can be derived for this process. It can also be shown that of other sugars that enter the cell, some act as competitive inhibitors toward sorbose uptake, suggesting that a group of substrates shares a common permease.

Much painstaking work may be summarized by saying that the chloroplast in general can supply ATP to the cytoplasm and that, in general, the chloroplast envelope is impermeable to ATP. Evidence that the chloroplast exports energy comes from several experiments. First, the work of MacRobbie (1965) showed that potassium ions were accumulated by certain algae, that ATP was required for the process, and that it was inhibited by cyanide (which stops the production of ATP by mitochondria) in the dark but not in the light. Hence the chloroplast supplied the ATP. Tanner and coworkers (1966) showed that glucose uptake in *Chlorella* depended on ATP that could be supplied by cyclic photophosphorylation. Secondly, other cell activities can be seen, such as protoplasmic streaming, which can be shown to be ATP-dependent. It is fastest in bright light, and the most effective wavelengths are correlated with (cyclic) photophosphorylation. Thirdly, as pointed out by Heber (1974), the ratio of the concentrations of ATP and ADP is higher outside than inside chloroplasts, although in light the overall rate of respiratory oxidation of NADH (and therefore of respiratory ATP production) falls, so that the ATP accumulation in the cytoplasm originates from the chloroplast.

The impermeability of the chloroplast envelope to ATP has been debated by many workers (reviewed by Walker, 1974). A convincing experiment showed

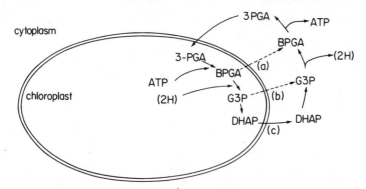

Figure 10.1. Three stages (a, b, c) in the development of the concept of a shuttle system for export of ATP from the chloroplast. Note that (2H) is also exported by (b) and (c) (see Figure 10.2)

that isolated and intact chloroplasts in which ATP production had been artificially inhibited (thus inhibiting photosynthesis) did not perform PGA-dependent oxygen evolution even when ATP was added to the medium, until the envelope was disrupted, implying that the failure to carry out the reaction was due to the lack of penetration of ATP (Stokes and Walker, 1971). (The inhibitors used were not expected to affect directly any possible ATP translocation.) Heldt and coworkers (1974) has reviewed studies of adenylate uptake by intact chloroplasts in which it was found that there was a carrier for ATP and ADP, but it was very slow. Export of ATP from the chloroplast must therefore be by indirect means.

The obvious indirect method is for the chloroplast to export carbohydrate which is then respired by the mitochondria. If this process took place it would require oxygen and be inhibited by cyanide, contrary to observation. Another possibility would be for the chloroplast to export 1,3-BPGA*, which might then regenerate 3-PGA and ATP in the cytoplasm, the PGA being taken up again by the chloroplast to complete a cycle. Such a cycle, of which the net effect is to transport a metabolite from one side of a biological membrane to the other may

* See the footnote on page 77.

be termed a shuttle (Figure 10.1(a)). This particular shuttle is discounted because BPGA is not able to cross the envelope at a reasonable speed, nor is it found in high enough concentration to be a credible shuttle-substrate. However, if the cycle is enlarged (b) by the addition of triose-phosphate dehydrogenase on each side (NADP-coupled inside, NAD-coupled outside) then these objections are overcome; the scheme is improved still further by recognition of the equilibrium between dihydroxyacetone phosphate and glyceraldehyde phosphate (greatly in favour of the first) so that triosephosphate isomerase is included too. The question, 'How does ATP cross the chloroplast envelope?', can be answered with some confidence in terms of the shuttle of Figure 10.1(c).

This shuttle also exports reducing equivalents together with the ATP. These are not required in the cytoplasm and some attention has been given to further

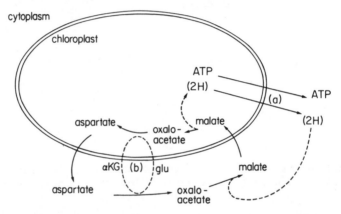

Figure 10.2. Elaboration of Figure 10.1 to show the recovery by the chloroplast of (2H) exported with ATP. (a) represents the shuttle of Figure 10.1. (b) provides for recovery of nitrogen by means of a third shuttle, using glutamate (glu) and α-ketoglutarate (αKG). (after U. Heber, 1974)

elaborations of the shuttle which could account for their return. Several authors have used the abundance of malate dehydrogenase inside and out to write a shuttle in which external NADH reduces oxaloacetate to malate. Malate is known to be rapidly translocated and to be oxidized back to oxaloacetate inside the chloroplast. The only difficulty, pointed out by Heber, is that the concentration of oxaloacetate is very low with respect to the Michaelis constant of the translocator, and therefore such a shuttle could not operate at the required speed. Heber (1974) proposed incorporating transamination steps so that oxaloacetate was transformed to aspartate, which answered that problem. The reverse flow of nitrogen was provided by the couple α-ketoglutarate and glutamic acid, as shown in Figure 10.2.

The evelope is of course permeable to some form of carbon dioxide. Bicarbonate ions are discounted here since carbonic anhydrase is almost entirely located inside the chloroplast and also the concentration ratio of bicarbonate inside and outside the chloroplast was found by Heldt to be proportional to the pH difference. Thirdly, entry of a charged species (HCO_3^-) would require co-transport of hydrogen ions to maintain pH and electric balance. The envelope is almost completely impermeable to hydrogen ions. Hence the membrane is likely to be specifically permeable to CO_2, probably by diffusion rather than by a permease.

Heldt and coworkers (1974) has shown that the dicarboxylic acids malate, aspartate, glutamate, α-ketoglutarate and oxaloacetate are translocated by the same carrier, and show mutual competitive inhibition. In the same way there is a specific translocator for inorganic phosphate, DHAP and PGA. This allows a shuttle to preserve the intra-chloroplast phosphate content, which would otherwise by diminished by the export of phosphate derivatives of carbohydrates (see Chapter 5). The same shuttle also forms part of the multiple shuttle-system shown in Figure 10.2.

Contrary to the facilitated systems for anion transport, the envelope is relatively impermeable to H^+, Na^+, Mg^{2+} and K^+. An effect of this must be that

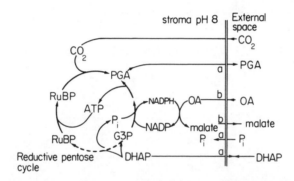

Figure 10.3. Use of shuttles (a and b) to achieve fixation of CO_2 in the dark, in isolated chloroplasts. (From Werden and coworkers, 1975, with permission)

the ion movements that accompany energy-conserving electron transport by thylakoids (uptake of H^+, followed more slowly by release of Mg^{2+}, must alter appreciably the pH and ionic composition of the stroma. This makes necessary a study of the enzymes of the stroma, since these factors may activate or regulate in some other way the uptake and fixation of carbon dioxide.

Werden and coworkers (1975) have demonstrated that the stroma enzymes may indeed be regulated by pH. Chloroplasts could be induced to fix carbon dioxide in the dark, given a source of ATP (in the form of triose phosphate, DHAP), provided the pH of the suspension was above 8. In this experiment (Figure 10.3) oxaloacetate was added in order to remove reducing equivalents not required by the fixation process. There is a clear indication of the importance of hydrogen ions, although it is still open to question whether the effect is direct or mediated via another ion such as Mg^{2+}. The use of the shuttle systems outlined above makes an interesting example of compartmentation in cell metabolism.

10.2 Carbon metabolism

10.21 RuBP carboxylase and glycollate

The enzyme RuBP carboxylase ('Fraction I protein') that incorporates carbon dioxide into PGA in the reductive pentose cycle (see section 5.1) presented a problem for many years in that although it is enormously abundant (estimated at 30% of the total leaf protein) the total activity appeared to be low and did not account for the rate at which carbon fixation could occur under optimum field conditions. This was a weakness in the theory of the pentose cycle. When the purified enzyme was studied kinetically, it was found that the Michaelis constant for carbon dioxide (the concentration of carbon dioxide giving a rate half that of the rate at saturating concentrations of all reactants) was approximately 600 μM. This is more than the concentration of carbon dioxide in air-saturated water (8–10 μM). The carbon dioxide concentration will be even lower in the chloroplast because of diffusion resistance, and so the enzyme could, it seemed, only ever work at a small fraction of its calculated maximum rate.

Table 10.1. The multiple forms of RuBP carboxylase

Kinetic form	Apparent K_m for HCO_3 (mM)	Calculated K_m for CO_2 (μM)	Relative V_{max}	Activity at 9·0 μM CO_2
Low-K_m	0·5–0·8	13–22	1·0	24
Intermediate—K_m	2·5–3·0	67–81	3·5–4·0	27
High-K_m	20–25	540–670	0·8–0·9	1
Intact chloroplasts	0·6	25		20

Assays performed at pH 7·8; in the presence of 10 mM $MgCl_2$ and either R5P or FBP to stabilize the low-K_m form. From Jensen and Bahr (1974).

More recent work showed that this enzyme had some unexpected features, one of the more striking of which was the need to keep it at a high concentration of total protein (which necessitated a rigorous technique for assay of RuBP carboxylase activity). Jensen and Bahr (1974) reported comparative kinetic data which are set out in Table 10.1. The enzyme could exist in one of three possible states, which possessed different Michaelis constants for carbon dioxide, the lowest of which (13–22 μM) removed the previous doubt affecting the validity of the pentose cycle.

Another recent discovery opened up a wide field of discussion and drew together several phenomena, known for many years, that had presented a number of problems. This was the announcement by Bowes, Ogren and Hageman (1971) that the purified RuBP carboxylase was found to have the property of a second type of reaction. Instead of cleaving the RuBP substrate by the incorporation of carbon dioxide, it could insert a molecule of oxygen so as to produce one molecule each of phosphoglycollate and PGA according to the scheme:

$$
O_2 \;+\;
\begin{array}{l}
CH_2O(P) \\
| \\
CO \\
| \\
CH.OH \\
| \\
CH.OH \\
| \\
CH_2O(P)
\end{array}
\quad\xrightarrow[\text{(oxygenase reaction)}]{\text{RuBP carboxylase}}\quad
\begin{array}{l}
CH_2O(P) \\
| \\
COOH
\end{array}
\;+\;
\begin{array}{l}
COOH \\
| \\
CH.OH \\
| \\
CH_2.O(P)
\end{array}
$$

  oxygen RuBP   phosphoglycollate   PGA

which may be compared with reaction 1 of the reductive pentose cycle (p. 81). So far, all samples of the enzyme possess the two types of activity, and there has been no successful separation of the two by improved purification procedures. Considerable attention has been paid to the comparative kinetics of the reactions, however, and it does appear that the relative rates may vary widely. An important factor pointed out by Tolbert and Ryan (1974) was the mutual competitive inhibition between oxygen and carbon dioxide in each reaction. Hence it would follow that phosphoglycollate should be more readily formed at high relative pressures of oxygen. In fact phosphoglycollate is observed only with difficulty, since there is an active phosphatase that liberates glycollate. Since the pioneering work of Zelitch (see Zelitch, 1971) in this field (mainly on tobacco, *Nicotiana*) it was known that glycollate was indeed produced more markedly when the carbon dioxide concentration was low relative to oxygen. Whittingham reported the accumulation of glycollate in the culture medium of algae such as *Chlorella*, and both *Nicotiana* and *Chlorella* were limited in their growth rates by this tendency to form glycollate.

It is worth noting that the origin of glycollate, which may seem obvious and simple in the oxygenase reaction described above, in fact taxed many minds and generated many ideas now considered false. Such hypotheses included a leakage from the transketolase reaction (p. 83) so that the C_2-fragment carried by thiamine pyrophosphate became detached; a reductive condensation of two carbon dioxide molecules, and reduction of acetate and formation via an independent source of glyoxylate. Tolbert and Ryan (1974) put these clearly in perspective. To be acceptable, any scheme has to be a demonstrable biochemical reaction, to explain the prior formation of the phosphate ester (phosphoglycollate), to account for the appearance of ^{18}O from isotopically-labelled oxygen (O_2) in the carboxyl group of the glycollate, as well as the observations concerning dependence on a high O_2/CO_2 ratio (including the effect of high pH).

The further metabolism of glycollate was substantially worked out prior to the discovery of its means of formation, in the laboratories of Zelitch, Whittingham and Tolbert. Since the pathway evolves carbon dioxide, it accounts for one half of the process of photorespiration and has been extensively studied, since it acts so as to counter carbon dioxide fixation in photosynthesis. The pathway involves different regions of the cell, particularly the chloroplast, cytosol, mitochondria and microbodies. These organelles are all shown in Plate 1 (p. 12). One key to

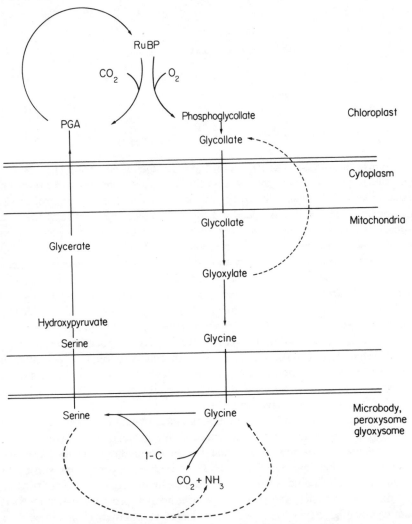

Figure 10.4. Metabolic pathway for recovering glycollate lost by the oxygenase reaction. Note the involvement of various cytoplasmic compartments. (After Tolbert, 1971)

this pathway was the radio-isotopic labelling of the intermediate L-serine. This was found to have all its carbon atoms equally labelled under conditions of high photorespiratory activity, and to be chiefly carboxyl-labelled when photorespiration was minimal. The production of serine from photosynthesis is set out in Figure 10.4 which is an abbreviated version of a scheme set out by Tolbert

(1971). There are a number of points to be made. First, there is represented (by a dotted line) the re-entry of glyoxylate into the chloroplast where it can be reduced back to glycollate by means of an NADPH-specific reductase. This would provide a glycollate-dependent NADPH-oxidase cycle, and thus consume molecular oxygen and reducing power from photosynthesis (see Zelitch, 1971). However, it does not account for the production of carbon dioxide, which arises from the conversion of glycine to serine, using the (1-C) group derived from the decomposition of a second molecule of glycine (not glyoxylate). Tolbert has doubted whether this is sufficient to account for all photorespiratory carbon dioxide;* however, pathways are known (particularly in animal tissues) which oxidatively convert serine (via choline and betaine) to glycine, carbon dioxide and ammonia. Since the diagram involves the passage into and out of the field of nitrogen metabolism, it should be noted that the formation of glycine from glyoxylate is by transamination from an amino acid; since the nitrogen is released from glycine as ammonia, there is a need for the re-incorporation of ammonia into amino acids which requires ATP. Lastly, the function of the overall pathway is still mysterious. Serine can be made from PGA (upper left, Figure 10.4) with the loss of one equivalent of ATP, so if the reactions from phosphoglycollate to glycine were eliminated there would be no real loss of metabolic capability. We may tentatively conclude that evolution has not found a way of eliminating the oxygenase reaction from the RuBP carboxylase enzyme and that the glycollate pathway exists as a scavenger mechanism that recovers three out of four carbon atoms.

10.22 Photorespiration

Photorespiration, defined as a light-dependent uptake of oxygen and output of carbon dioxide, can be seen to have a good biochemical basis in the processes described above. The key is glycollate; the formation of glycollate, and its subsequent oxidation to glyoxylate, account for the uptake of oxygen; and the carbon dioxide is evolved during the recovery of PGA from two molecules of glyoxylate. The dependence on light arises partly because the RuBP carboxylase is activated by the high pH of the stroma in the light and by magnesium ions (among other agents) which are released from the thylakoids into the stroma during illumination, and partly because the substrate RuBP cannot be regenerated from PGA without ATP and NADPH, produced by thylakoids in light.

If the equation of photosynthesis is written as an equilibrium

$$CO_2 + H_2O \rightleftharpoons (C(H_2O)) + O_2$$

then photorespiration is represented by the right-to-left direction. The direction of overall change can be seen to be dependent, at least in part, on the relative partial pressures of the two gases, oxygen and carbon dioxide, and, as predicted by

---

* Labelling studies with $^{18}O_2$ in algae show that some of the photorespiratory CO_2 is labelled in both oxygen atoms which could not arise by the usually accepted form of the glycollate pathway. In soybeans, however, $C(^{18}O_2)$ was only rarely observed (Gerster and coworkers, 1974).

186

Le Chatelier's principle, plants in conditions of light and a high concentration ratio of oxygen to carbon dioxide may lose carbon to the atmosphere by way of photorespiration. This also provides a rationale for the observation of Warburg (1920) that photosynthesis is inhibited by oxygen (the Warburg Effect, discussed by Rabinowitch, 1945b).

Conditions under which the rates of the photosynthetic and respiratory reactions balance each other are termed *compensation points*. The carbon dioxide compensation point for many temperate species is in the range 50–100 ppm,

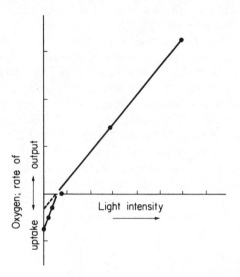

Figure 10.5. The Kok Effect. The apparent increase in photosynthetic efficiency at light intensities below the compensation point. *Chlorella* cells, in a Warburg manometer. (From Kok, 1951, with permission)

depending on the intensity of illumination and the pressure of oxygen. Note that the concentration of carbon dioxide in air is 300 ppm. At low light intensities respiration outweighs photosynthetic gas exchange, so that there is also a compensation point in the intensity-dependence curve, and it depends on the oxygen to carbon dioxide ratio. However, the light-variable component of the oxygen uptake (photorespiration) is not directly proportional to intensity, as shown by Hoch and Owens (1963) who used isotopically-labelled oxygen to separate the two rates. This provided an explanation for the Kok Effect (Figure 10.5) in which the quantum yield of photosynthesis appears to be twice as great below the compensation point than above it when measured as a function of light-intensity. At the greater intensities some quanta operate photorespiration, thus cancelling the effect of others.

The practical application of our knowledge of photorespiration has caused some concern. It might be possible to explore any natural variation in the enzymological characteristics of the enzyme by genetics and plant breeding, and hence to enable agriculture to produce useful yields under conditions where at present photorespiration sets a limit. A more obvious practical approach is to cultivate plants (necessarily high cash-value plants such as unseasonal tomatoes and flowers) in enclosed greenhouses where the atmosphere can be relatively enriched in carbon dioxide over oxygen. In practical terms, however, other limits may operate, such as the resistance to diffusion of carbon dioxide and of oxygen from the chloroplasts through cell walls, intercellular air-spaces and stomata. This often results in the concentrations of the gases at the point of metabolism being very different from the atmospheric proportions. Furthermore, the resistance to diffusion is affected by the stomata, which open and close both in response to light and water stress, and other factors which are characteristic of each species. Depending on the anatomical form of the leaf, a relatively large proportion of the water lost by the leaf passes through the stomata, so that under conditions of water stress (or even at set times of day when water stress might be expected) the stomata may close and restrict photosynthesis.

From the foregoing it will be apparent that great advantage will accure to a plant which can arrange a high carbon dioxide concentration relative to oxygen at the point of metabolism, or which can prevent the closing of its stomata from reducing its rate of photosynthesis. Such plants exist and form the subject of the next section.

10.23 The C_4 pathway of carbon dioxide fixation

Within the last few years there has been a rebirth of interest in the metabolism of carbon dioxide, resulting in the formulation of a new pathway. This pathway is usually known as the C_4 pathway, because of the key observation that, in sugar cane, a pulse of radioactive carbon dioxide appeared most rapidly, not in PGA (see p. 77) as in *Chlorella* and most temperate-zone vegetation, but in compounds such as malic, oxaloacetic and aspartic acids, which have four carbon atoms (Kortschak, Hartt and Burr, 1965). There is nothing extraordinary in this, since many tissues, including mammalian liver, can fix carbon dioxide into C_4-acids. However, there was a difference in that in sugar cane the process was light-dependent and appeared to lead to entry of the newly fixed carbon into the general metabolic pool, resulting in sucrose synthesis. (In non-photosynthetic tissues, although the carboxylation of phosphoenolpyruvic acid to oxaloacetic acid is of vital importance, it clearly leads nowhere permanent and the carbon dioxide must eventually be released again by what is in effect reversal of the reaction by which it was fixed.) Hatch and Slack (1966) pointed out that in tropical grasses (of which sugar cane is a spectacular member) uptake in light of carbon dioxide from a closed space continues until the gas becomes virtually undetectable (0·10 ppm), whereas the usual (temperate) pattern is for the level to fall only as far as the compensation point of about 60 ppm where photosynthetic uptake is balanced by respiratory release of carbon dioxide. It was clear that the tropical grasses contained an enzyme system capable of capturing carbon dioxide at much lower concentrations than those effective with the RuBP carboxylase.

The simplest representation of the C_4 pathway is a cycle (Figure 10.6) of five enzymes:

(a) pyruvate, orthophosphate, dikinase (EC.2.7.9.1)

$$\text{ATP} + \underset{\substack{| \\ CH_3}}{\overset{\substack{COOH \\ |}}{CO}} + P_i \longrightarrow \underset{\substack{\| \\ CH_2}}{\overset{\substack{COOH \\ |}}{C}} - O\textcircled{P} + PP_i + AMP$$

$$\text{pyruvate} \qquad\qquad \text{phosphoenolpyruvate (PEP)}$$

(b) phosphoenolpyruvate carboxylase (EC.4.1.1.31)

$$\underset{\substack{\| \\ CH_2}}{\overset{\substack{COOH \\ |}}{C}} - O\textcircled{P} + CO_2 \longrightarrow \underset{\substack{| \\ CH_2 - COOH}}{\overset{\substack{COOH \\ |}}{CO}} + P_i$$

$$\text{oxaloacetate}$$

(c and d) malate dehydrogenases, NADP- and NAD-linked respectively, EC.1.1.1.82 and EC.1.1.1.37

$$\underset{\substack{| \\ CH_2.COOH}}{\overset{\substack{COOH \\ |}}{CO}} + NAD(P)H + H^+ \longrightarrow \underset{\substack{| \\ CH_2.COOH}}{\overset{\substack{COOH \\ |}}{HCOH}} + NAD(P)^+$$

$$\text{L-malate}$$

and (e) the malic enzyme (L-malate dehydrogenase, decarboxylating, NADP), EC.1.1.1.40

$$\underset{\substack{| \\ CH_2.COOH}}{\overset{\substack{HCOH.COOH \\ |}}{}} + NADP^+ \longrightarrow \underset{\substack{| \\ CH_3}}{\overset{\substack{CO.COOH \\ |}}{}} + CO_2 + NADPH + H^+$$

$$\text{pyruvate}$$

Reaction (a) is specific to this pathway. It is remarkable because the free energy of hydrolysis of PEP is usually quoted as $58 \cdot 6$ kJ mole^{-1} (14 kcal mol^{-1}), which makes the enzyme pyruvate kinase (in the glycolysis pathway) virtually irreversible in the direction of ATP synthesis. The difference here may be first that ATP is balanced at a higher value of the quotient $[\text{ATP}]/[\text{ADP}][P_i]$, so that the actual free energy of hydrolysis of ATP is raised above its standard value of $30 \cdot 5$ kJ mole^{-1} ($7 \cdot 3$ kcal mol^{-1}). Secondly, there is a powerful pyrophosphatase which removes the pyrophosphate formed as shown in the equation, so that in effect ATP loses two 'high energy' phosphate units. Enzyme (b) is widespread;

an interesting point is that carbon dioxide reacts in the f[...]
compared to RuBP carboxylase which uses CO_2.* This all[...]
the C_4 pathway to discriminate against the isotope ^{13}C wh[...]
extent in natural carbon dioxide. Hence C_4 and C_3 plants[...]
with a mass spectrometer. Enzyme (c) is a chloroplast-s[...]

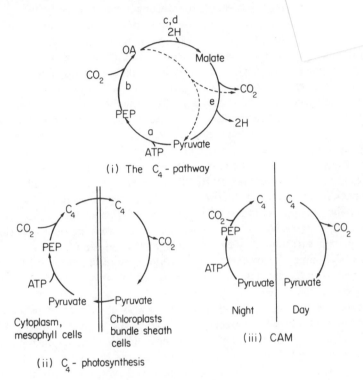

(i) The C_4 - pathway

(ii) C_4 - photosynthesis

(iii) CAM

Figure 10.6. Cyclic arrangement of the enzymes of the C_4 pathway and the arrangements for its operation in C_4-photosynthesis and Crassulacean acid metabolism (CAM) respectively

ubiquitous enzyme (d). (c) is light-activated (Johnson and Hatch, 1970). (e) is widely distributed.

The arrangement of these enzymes into a cycle at first sight merely creates a pyruvate-dependent ATPase system. To make sense of an arrangement such as that of Figure 10.6(i) either a spatial or temporal separation of the enzymes must be envisaged.

In the first case (Figure 10.6(ii)) the cycle operates so as to produce a flow of carbon dioxide from left to right. This flow is coupled to the hydrolysis of ATP so that the carbon dioxide may be thought of as pumped by the cycle. This gains

* The enzyme carbonic anhydrase catalyses the equilibrium $CO_2 + OH^- = HCO^-_3$. It is surprisingly very abundant in C_3 plants, which use CO_2 and should not need the enzyme, and present only in the cytoplasm of C_4 plants. Everson (1971) suggests a buffering role for the CO_2/HCO^-_3 system, against the sudden pH fluxes due to the ion-pumping of the thylakoid system.

point from the intercellular and intracellular compartmentation of the enzymes of the cycle. Carbon dioxide at low concentration is rapidly fixed, giving oxaloacetate, malate or aspartate at a much higher concentration. The C_4 acids pass by diffusion (or possibly active uptake) into chloroplasts of the same or different cells, and are there decomposed releasing carbon dioxide and a C_3 compound which returns as pyruvate to the starting point completing the cycle. Carbon dioxide is released at the point of metabolism at a concentration much higher than atmospheric, so that the carboxylase activity of RuBP carboxylase has a great advantage over the oxygenase activity. In fact PEP carboxylase which is found in the cells bordering the air spaces inside the leaf (mesophyll cell) is able to bring about an almost complete depletion of carbon dioxide from the air spaces, which must speed the inward diffusion of the gas from the atmosphere. At the same time it explains the very low compensation point for carbon dioxide in C_4 plants, and also their apparent lack of photorespiration (we shall see that in some cases the oxygen concentration can be diminished).

For the C_4 pathway to be effective the carbon dioxide released from the C_4 carrier at the point of uptake by RuBP carboxylase must not be allowed to escape back into the air spaces. This can in principle be achieved in two ways. The first is the obvious circumstance where the C_4 molecule passes from the cytoplasm, where it was made, to the chloroplast within the same cell; since the chloroplast is surrounded by cytoplasm any back-diffusion will be counteracted by the 'CO_2 pump'. However, the situation is more complex. All C_4 species examined so far display a type of anatomy termed *Kranz* (crown) (see Plate 12) in which the vascular bundles (the strands of cells adapted to transport water and nutrients) are surrounded by close layers of cells containing chloroplasts (the bundle-sheath), with no air spaces. The sheath is surrounded by mesophyll cells usually making good contact with the bundle-sheath, although this may not be complete; notice the air space in contact with the bundle-sheath in Plate 12. Although the bundle-sheath cell may have some contact with an air space Laetsch (1974) suggests that the bulk of CO_2 has to pass in via the mesophyll cells, so that the CO_2 in the bundle-sheath is effectively protected by many layers of cytoplasm and cell walls.

Several groups of workers (see, for example, Black (1974)) have developed methods for dissecting these leaves so that the enzyme complements of bundle-sheath and mesophyll cells can be compared. In one major group of the C_4 plants, the mesophyll cells contain nearly 100 times as much PEP carboxylase as the bundle-sheath cells, while for the 'malic enzyme' that liberates carbon dioxide from malate the bundle-sheath contains a similarly high proportion with respect to the mesophyll. Therefore, in this group, there is a strong presumption that malate is synthesized in the mesophyll and (although a positive proof is lacking) transported to the bundle-sheath cells. The malic enzyme here uses the coenzyme NADP. Another group possesses a malic enzyme specific for NAD and a third group decarboxylates oxaloacetate by means of the enzyme PEP carboxykinase, regenerating PEP. These last two groups contain high levels of transaminase which converts oxaloacetate reversibly to aspartate, in both types of cell. The

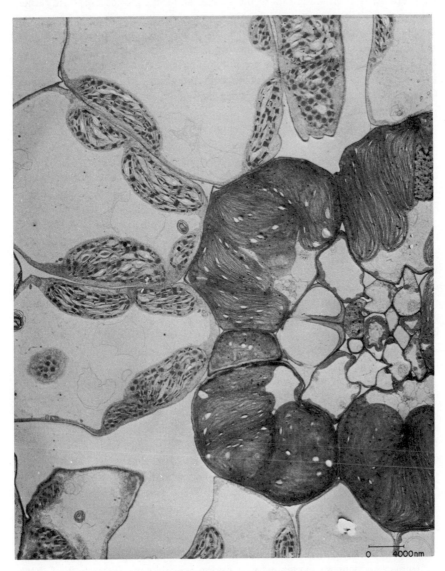

Plate 12. 'Kranz' anatomy characteristic of C_4 plants. Micrograph of a crabgrass (*Digitaria sanguinalis* (L.) Scop.) leaf cross-section the dimorphism between bundle-sheath and mesophyll chloroplasts. Grana in the bundle-sheath are only rudimentary, whereas they are prominent in the mesophyll. Starch grains are present in both. Magnification ×4000. From C. C. Black, Plant Physiology (1971), **47,** pp. 15–23, with permission of the publisher.

suggestion has been made that in the second and third groups aspartate rather than malate is the C_4 carrier. In the first group pyruvate is returned from the bundle-sheath to the mesophyll cells, in the second it is probably alanine, and in the third either pyruvate or alanine (or even, it has been suggested, PEP); however, phosphate compounds are not commonly diffusible across cell membranes, a point which may be answered by the abundance of plasmodesmata (strands of continuous cytoplasm connecting cells), which might well allow such a passage to take place.

The mesophyll chloroplasts of all the groups appear to be reasonably similar. They possess grana, carry out normal electron transport and generate ATP. They do not have much ability to fix carbon dioxide by the reductive pentose cycle. They contain the enzymes for generating PEP and for reducing oxaloacetate to malate with NADPH; however, the enzyme PEP carboxylase is located in the cytoplasm. The bundle-sheath chloroplasts on the other hand, show more variety. At one extreme is the dimorphism shown by *Sorghum*: the bundle-sheath chloroplasts are often devoid of grana and patently different from mesophyll chloroplasts in size, shape and electron-microscopic appearance. A wide range of tests has failed to detect any activity of photosystem II, and non-cyclic electron transport does not appear to take place, although cyclic phosphorylation has been observed. At the other extreme bundle-sheath chloroplasts may show no difference in structure (in which case they are often bigger) or in photochemical activity. However, nearly all reports indicate that RuBP carboxylase is much more abundant in the sheath than in the mesophyll chloroplasts.

In the case of plants showing a great degree of dimorphism, it is suggested (Hatch and Slack, and others) that malate enters the bundle-sheath chloroplasts and gives rise to carbon dioxide, NADPH and pyruvate; RuBP carboxylase assimilates the carbon dioxide to give PGA, and half of this is reduced with the NADPH (the chloroplast does not produce NADPH from non-cyclic electron transport) and forms carbohydrate by the reductive pentose cycle. The other half of the PGA returns to the mesophyll together with the pyruvate. The PGA is reduced to carbohydrate by the abundant PGA kinase and triosephosphate dehydrogenase. The carbohydrate then returns to the bundle-sheath and is stored as starch. This scheme could hardly be more difficult for the observer, whether or not it is so for the plant (see Laetsch (1974) for criticism). A simplifying minority view (Bucke and Coombs, 1974) would have all the C_4 and C_3 (reductive pentose cycle) activity taking place in the mesophyll of these dimorphic plants, transporting only sucrose to the bundle-sheath, where it is laid down as starch. In this view the bundle-sheath chloroplast makes virtually no use of its chlorophyll but acts as an amyloplast.

In non-dimorphic plants there appears to be a relatively straightforward division of activity—the mesophyll generates malate in its chloroplasts, or aspartate in the cytoplasm, which then move to the bundle-sheath chloroplasts where the carbon dioxide is released and converted to starch by the conventional reductive pentose cycle (Figure 5.1).

The value of the C_4 operation (according to the foregoing hypotheses) is the same in both types of plant—that carbon dioxide at the point of fixation in the stroma of the bundle-sheath chloroplasts is at a much higher concentration than in the C_3 plants. This in turn results in (i) the velocity of the RuBP–carboxylase-catalyzed reaction being closer to the saturating rate, (ii) successful competitive inhibition, by the high carbon dioxide concentration, of the oxygenase reaction of the same enzyme (assisted by the lack of oxygen generation in the bundle-sheath of dimorphic species), (iii) the effective scavenging of carbon dioxide from the sub-stomatal air spaces so that its inward diffusion is facilitated with respect to loss of water vapour and (iv) the carbon dioxide arising from photorespiration in non-dimorphic plants being retained in the C_4–C_3 cycle so that it can not diffuse to an air space and be lost. All this is achieved at a cost of energy. In the pathway represented in Figure 10.6 each molecule of carbon dioxide is photosynthesized twice, roughly doubling the energy requirement. If the maximum utilization of absorbed solar energy by a C_3 species is 27% then that of C_4 species is probably approximately half that, i.e. 14%, a figure which is still very much higher than is actually achieved by any field crop. Since the limiting factor in photosynthesis is usually the rate of the RuBP carboxylase reaction, the foregoing explains why C_4 plants outperform C_3 plants under natural conditions, but when carbon dioxide levels are raised and oxygen levels diminished C_3 plants can approach or even overtake C_4.

10.24 Crassulacean acid metabolism (CAM)

Essentially the C_4 pathway of the preceding section was a CO_2-pump transferring the gas from one locality to another and increasing its partial pressure. Many succulent plants make use of the same enzymes to carry out a temporal transfer rather than a spatial one, that is, a transfer from *night* when the plant opens its stomata and takes in carbon dioxide to *day* when the plant conserves water with closed stomata (Figure 10.6(iii)). This is achieved in the very abundant mesophyll (which gives the succulents their characteristic fleshy texture); during the night oxaloacetate, malate and aspartate accumulate in the vacuoles of the mesophyll cells, so that in the morning the leaf is notably sour in taste. During the day the acidity diminishes and the carbon dioxide is released from the C_4-acids and re-incorporated via the reductive pentose cycle. The principal difference between CAM and C_4 plants (apart from the anatomy) would be that the former form the C_4-acids in the dark (presumably using ATP from mitochondria) whereas the latter almost certainly use photosynthetically-generated ATP and NADPH.

There appears to be little natural classification in the C_4-habit. Downton (1975) has listed thirteen families of angiosperms (117 genera) having C_4 species (485). The monocotyledons are represented by the Cyperaceae and the Gramineae, which latter is the most numerous, containing the well known genera maize (*Zea*), *Sorghum*, sugar-cane (*Saccharum*) and millet (*Eleusine*). In the dicotyledons, the families represented are the Azioaceae, Amaranthaceae

(*Amaranthus*), Chenopodiaceae (*Atriplex*), Compositeae, Euphorbiaceae (*Euphorbia*), Myctaginaceae, Portulacaceae and the Zygophyllaceae. There are many members of these families that do not show C_4 photosynthesis. Smith and Robbins (1974) describe some evolutionary relationships between members of subdivisions of the above families, and conclude that in many groups the C_4 habit has appeared independently. They also argue that CAM is, like C_4 photosynthesis, an evolutionary modification of the universal C_3 system and arose independently; for example, the genus *Euphorbia* contains C_3, C_4 and CAM species.

10.3 Chloroplast autonomy

The chloroplast possess its own DNA, as does the mitochondrion. The question, therefore, arises, 'to what extent is the growth of the organelle controlled from within rather than from the DNA in the cell nucleus?'. The chloroplast DNA is in many cells distinguishable from nuclear DNA in that the former has a greater density, is circular rather than open-ended, and is not associated with protein. All these characteristics are shared with prokaryotic organisms (the bacteria and the blue-green algae). In addition, the chloroplast has another prokaryotic feature in that the ribosomes have a sedimentation coefficient of 70S; cytoplasmic ribosomes on the other hand are 80S. For discrimination between 80S and 70S ribosomes it has been found, with consistent results to date, that 80S ribosomes are inhibited by cycloheximide, and 70S ribosomes by the antibiotics lincomycin, chloramphenicol and streptomycin.

Ridley and Leech (1970) cultured chloroplasts in cell-free media, for several days, during which time a division took place. No growth was observed.* Since the medium contained all the materials that should have been required for protein synthesis to take place, why did the organelles fail to increase in size? The answer appears to lie at least in part in the observation that many proteins in the chloroplast are controlled from the cell nucleus. This is the case for the many known mutants that block the formation of chlorophyll or carotenoid pigments, and the synthesis of protein carriers in the electron-transport pathway.

The inheritance of the mutation follows the laws of Mendel. One test which is often used is to correlate inheritance with the sex of the parent carrying the gene. If the gene is always transmitted from the female parent there is a presumption that it is coded on extra-nuclear DNA (in this case, the DNA of the chloroplast). So far only two proteins have been found which are maternally transmitted and which are believed to be coded by the chloroplast DNA. This is surprising since, although the quantity of DNA per chloroplast in various species varies between 10^{-16} and 10^{-14} (Kirk, 1966), and although there may be several copies of the

* There is an intriguing study by Taylor (1971) of a symbiotic situation in which chloroplast eaten by marine slugs (feeding on *Codium tomentosum* Stackh., a siphonalean seaweed) are transferred intact to the cells of the digestive glands and not only survive but can in the light successfully reverse the oxygen uptake of the slugs, making a major contribution to the animals' energy requirements as well as to their camouflage. This must be mainly owed to the toughness of the chloroplast envelopes. The genome of the slug does not provide for rejection of the foreign organelle—what an interesting organism would result if it acquired the ability to culture them!

genome per plastid, it is expected that the chloroplast DNA should contain sufficient information for some 500 proteins of molecular weights averaging 27 000. The two proteins that are under discussion are the 'large subunit' of fraction I protein (RuBP carboxylase) and some of the proteins that constitute the 30S subunit of the chloroplast ribosome. This was shown by genetic analysis of *Euglena* and *Nicotiana* (tobacco) as described above in the laboratory of Wildman (see Chan and Wildman, 1972) and supported by Criddle and coworkers (1970), who found that the synthesis of the two kinds of subunit of fraction I protein were affected differently by chloramphenicol and cycloheximide, and by Blair and Ellis (1973) who found that radioactive amino acids were found in the larger subunit of fraction I protein alone of the soluble proteins when incubated with intact, isolated chloroplasts. (They did, however, observe the bulk of the activity to be incorporated into insoluble material.) Similar arguments apply to the 30S (chloroplast DNA) and 50S (nuclear DNA) fractions of chloroplast ribosomal protein, (see Bourque and Wildman, 1973). It is interesting, but unexplained, why in each case the chloroplast is competent to synthesize only one part of a complex macromolecular assembly. Perhaps the need to put two parts together provides a means of interrelating growth of chloroplasts with growth of the cell as a whole.

The chloroplast DNA is also believed to code for the production of the chloroplast ribosomal RNA, probably all of it (see Heizman, 1974). In this case there is the likelihood that production is controlled by the cytoplasmic synthesis of regulators such as the 'magic spots'—guanosine tetra- and pentaphosphates (ppGpp and pppGpp).

What of the other chloroplast proteins, which follow Mendelian inheritance and are presumed to be coded on the nuclear DNA? In the case of cytochrome-552 in *Euglena* and ferredoxin–NADP oxidoreductase in *Chlamydomonas* the proteins are synthesized on the chloroplast ribosomes, while the NADP-linked triosephosphate dehydrogenase is assembled on the cytoplasmic ribosomes. In both cases there is a transfer problem, since the chloroplast envelope has to be passed either by the messenger RNA that carries the code from the nuclear DNA to the chloroplast ribosomes, or by the protein itself. This data has been summarized for *Euglena*, with reference to the control of chloroplast development in light, by Schiff (1974). Schiff (1973) has also pointed out the similarity between a green plant cell and the situation one could imagine developing from the invasion and symbiosis of a prokaryote within a eukaryotic host cell.

Appendix: physical constants, formulae, etc.

	symbol	value
Velocity of light *in vacuo*	c	$2 \cdot 998 \times 10^8$ m sec^{-1}
Planck's constant	h	$\begin{cases} 6 \cdot 626 \times 10^{-27} \text{ erg sec} \\ 6 \cdot 626 \times 10^{-34} \text{ J sec} \end{cases}$
Coulomb (ampere sec)	C	$2 \cdot 998 \times 10^9$ e.s.u. $6 \cdot 24 \times 10^{18}$ electrons
electronic charge	e	$\begin{cases} 4 \cdot 803 \times 10^{-10} \text{ e.s.u.} \\ 1 \cdot 602 \times 10^{-19} \text{ C} \end{cases}$
electronvolt (unit of energy)	eV	$1 \cdot 602 \times 10^{-19}$ J equivalent to $23 \cdot 053$ cal mol^{-1} equivalent to photon of $1239 \cdot 5$ nm
calorie (defined)	cal	$4 \cdot 1841$ J
Joule (absolute)	J	$1 \cdot 0 \text{ CV} = 10^7$ erg $= 0 \cdot 2390$ cal
Avogadro's Number (molecules per mole)	N	$6 \cdot 0225 \times 10^{23}$
Absolute zero	$0°$K	$-273 \cdot 16°$C
Faraday constant	F	$96 \cdot 487$ C equiv.$^{-1}$
Gas constant	R	$8 \cdot 3143$ J deg.$^{-1}$ mol^{-1} $= 1 \cdot 987$ cal deg.$^{-1}$ mol^{-1} $= 0 \cdot 08206$ litre atm. deg.$^{-1}$ mol^{-1}

Solar constant (radiant energy entering the earth's atmosphere) $8 \cdot 3$ J min^{-1} cm^{-2} ($2 \cdot 0$ cal min^{-1} cm^{-2})

$RT \log_e = 5804 \log_{10}$ J mol^{-1} at $25°$; *add* 19 J per degree rise in temperature.

$\qquad = 1387 \log_{10}$ cal mol^{-1} at $25°$; *add* $4 \cdot 6$ cal per degree rise in temperature.

$(RT/nF) \log_e = (59 \cdot 16/n) \log_{10}$ mV at $25°$; *add* $0 \cdot 20$ mV per degree rise in temperature.

Wavelengths of hydrogen lines used in the calibration of spectro-photometers:

656·3 nm, 486·1 nm.

Equations for the estimation of chlorophyll extracted in 80% acetone:

$$C_a = 12\cdot7\, E^{1cm}_{663nm} - 2\cdot69\, E^{1cm}_{645nm}$$
$$C_b = 22\cdot9\, E^{1cm}_{645nm} - 4.68\, E^{1cm}_{663nm}$$

from Hill (1963). C_a and C_b are the concentrations of chlorophylls a and b respectively in mg/litre; E^{1cm}_{663nm} is the extinction (absorbance) (log I_0/I)   a 1 cm path-length sample at a wavelength of 663 nm.

Figure A.1 presents a nomogram for the rapid solution of the above equations: locate the extinction values on the appropriate axes (note that the

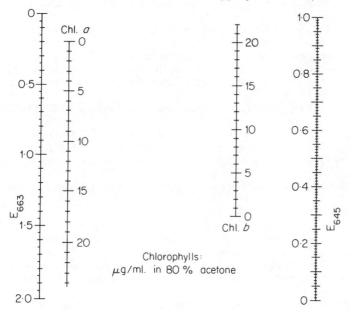

Figure A.1

direction of the scales are reversed) and join by a thread or a transparent ruler (but beware of refraction effects). The values of C_a and C_b may be read off from the points of intersection of the thread with the appropriate scales. N.B. For the estimation of chlorophyll in cells of algae such as *Scenedesmus*, acetone is not a good extracting solvent. Methanol is useful in some cases. Reference must be made to the literature for solvents and spectroscopic constants.

Answers to numerical problems

1. Once, on average, in 0·25 sec.
   One in $3·6 \times 10^8$ molecules.

2. (i) That the photosynthetic unit was of variable size (since the means of the groups were so far apart in terms of their standard deviations). The sizes were 272, 630, 1130, 2550, 5100 chlorophyll molecules per unit, respectively. These lie close to a doubling series (1, 2, 4, 8, 16), and suggest that larger units are formed as aggregates of small ones.
   (ii) In the mutant the same groups appear (with barely significant differences) but whereas the normal leaf had a preponderance of the 'classical' size unit (2550) the mutant showed the most common size to be 602.
   (iii) That the size of the photosynthetic unit may be under the control of developmental or photosynthetic factors. See 'plastic chlorophyll', pp. 101 and 119.
   Data composed from published work of Schmid and Gaffron (1968).

3. The energy of activation is $1·0$ eV, or $96·5$ kJ mole^{-1} (23 000 cal mole^{-1}). The energy of a 690 nm photon is $1·8$ eV, therefore the store has an energy level of $0·8$ eV (which is similar to the potential difference beween X and Y , see section 8.2).
   Data composed from work of Bertsch and Azzi, cited in Bertsch (1969).

4. (a) $69·3$ kJ mole^{-1} ($16·56$ kcal mole^{-1}).
   (b) $0·36$ V if 1 ATP was formed for 2 electrons passing.
   (c) Between water/oxygen ($E' = +815$ mV at pH 7) and Y_{II} (unknown); between PQA ($E_0 = 115$ mV *in ethanol*) and ferricyanide (if only system II is operating. If system I is included, ATP could be coupled to the span X_I to ferricyanide, or to PQA to P700). See section 8.2. *Problem by courtesy of Dr. A. R. Crofts.*

5. The shift was presumably due to either ATP, a high-energy precursor, or a chemiosmotic field (see p. 71). Uncoupling agents could be expected to prevent these from forming. In the Table, the values of the 530-nm effect are found to be linear when plotted against log [KCl] (equal to log $[K^+]_{out}$), suggesting that the effect is following an equation such as that for the potential ψ. The unknown $[K^+]_{in}$ which remains virtually constant is found from the graph where the 530-nm effect has zero magnitude, $\psi = 0$ and $[K^+]_{out} = [K^+]_{in} = 4·64 \times 10^{-5}$ M. Under illumination, when the effect has value $0·3$, the extrapolated concentration of KCl required to give an equivalent effect can be read from the graph: $[K^+]_{out} = 4·64$ M, and $\psi = 300$ mV, which is a reasonable value for the support of phosphorylation (see Chapter 9). *Problem by courtesy of Dr. A. R. Crofts.*

6. (i) In the case of hexose, derived from F6P, the label initially is predicted to be located in carbons 3 and 4, which should be equal in activity. During subsequent

turns the other atoms become labelled, but C3 and C4 remain the most active, and equally so.

(ii) The experimental data are in accord with the prediction that C3 and C4 would be the most active, but there is a slight but noticeable difference between the activities of C3 and C4 that requires additional explanation. *Problem composed from data of Gibbs (1963).*

7. 2×10^9.

8. 4210.

9. Three orders of magnitude (10^3).

10. Chlorophyll a: 0·78 mg/ml; b: 0·39 mg/ml
      rates:
  30·7 μmole O_2 mg^{-1} h^{-1} in absence of ADP ('basal rate'; state 4)
  61·9 in presence of ADP (state 3)
  75·4 in presence of NH_4Cl (fully uncoupled: indicates maximum rate of photochemical system)
  P/O ratio 1·5 (2·0 if the background rate is ignored; see section 9.2).

References

Anderson, J. M. (1975). *Biochem. Biophys. Acta,* **416,** 191–235.

Anderson, J. M. and R. P. Levine (1974). *Biochim. Biophys. Acta,* **357,** 118–126.

Anderson, J. M., D. C. Fork and J. Amesz (1966). *Biochem. Biophys. Res. Commun.,* **23,** 874–879.

Arnon, D. I. (1968). In T. P. Singer (Ed.), *Biological Oxidations,* Interscience, New York, pp. 123–170.

Arnon, D. I., H. Y. Tsujimoto and B. D. McSwain (1965). *Nature,* **207,** 1367–1572.

Avron, M. and G. Ben-Hayyim (1969). In H. Metzner (Ed.), *Progress in Photosynthesis Research,* Vol. 3, Institut für Chemische Pflanzenphysiologie, Tubingen, pp. 1185–1196.

Avron, M. and B. Chance (1966). In J. B. Thomas and J. C. Goedheer (Eds.), *Currents in Photosynthesis,* Donker, Rotterdam, pp. 455–464.

Babcock, G. T. and K. Sauer (1975). *Biochim. Biophys. Acta,* **376,** 329–344.

Baltscheffsky, H. (1974). In M. Avron (Ed.), *Proc. IIIrd. Int. Cong. Photosynthesis,* Elsevier, Amsterdam, pp. 799–806.

Baltscheffsky, M. and D. O. Hall (1974). *FEBS letters,* **39,** 345–348.

Barber, J., J. Mills and J. Nicolson (1974). *FEBS letters,* **49,** 106–109.

Bassham, J. A., A. A. Benson, D. L. Kay, A. Z. Harris, A. T. Wilson and M. Calvin (1954). *J. Amer Chem. Soc.,* **76,** 1760–1770.

Bendall, D. S. (1968). *Biochem. J.,* **109,** 46P–47P.

Bendall, D. S., H. E. Davenport and R. Hill (1971). *Methods Enzymol.,* **23,** 327–344.

Bensasson, R. and E. J. Land (1973). *Biochim. Biophys. Acta,* **325,** 175–181.

Bertsch, W. (1969). In H. Metzner (Ed.), *Progress in Photosynthesis Research,* Institut für Chemische Pflanzenphysiologie, Tubingen, pp. 996–1005.

Bertsch, W., J. West and R. Hill (1969). *Biochim. Biophys. Acta,* **172,** 525–538.

Black, C. C. (1966). *Biochim. Biophys. Acta,* **120,** 332–340.

Black, C. C. (1974). In M. Avron (Ed.), *Proc. IIIrd Int. Cong. on Photosynthesis,* Elsevier, Amsterdam, pp. 1210–1218.

Blair, G. E. and R. J. Ellis (1973). *Biochim. Biophys. Acta,* **319,** 223–234.

Blinks, L. R. (1957). In H. Gaffron and others (Eds.), *Research in Photosynthesis,* Interscience, New York, pp. 444–449.

Boardman, N. K. and J. M. Anderson (1964). *Nature,* **203,** 166–167.

Boardman, N. K., O. Bjorkman, J. M. Anderson, D. J. Goodchild and S. W. Thorne (1974). In M. Avron (Ed.), *Proc. IIIrd Int. Cong. on Photosynthesis,* Elsevier, Amsterdam, pp. 1809–1827.

Bothe, H. (1969). In H. Metzner (Ed.), *Progress in Photosynthesis Research,* Vol. 3, Institut für Chemische Pflanzenphysiologie, Tubingen, pp. 1483–1491.

Bourque, D. P. and S. G. Wildman (1973). *Biochem. Biophys. Res. Commun.,* **50,** 532–537.

Bowes, G., W. L. Ogren and R. H. Hagerman (1971). *Biochem. Biophys. Res. Commun.*, **45**, 716–722.

Boyer, P. D. (1965). In T. E. King, H. S. Mason and M. Morrison (Eds.)., *Oxidases and Related Redox Systems*, Vol. 2, Wiley, New York, pp. 994–1017.

Boyer, P. D. (1968). In T. P. Singer (Ed.), *Biological Oxidations*, Wiley–Interscience, New York, pp. 193–235.

Branton, D., S. Bullivant, N. B. Gilula, M. J. Karnovsky, H. Moor, K. Mühlethaler, D. H. Northcote, L. Packer, B. Satir, P. Satir, V. Speth, L. A. Staehelin, R. L. Steere and R. S. Weinstein (1975). *Science*, **190**, 54.

Breton, J. and P. Mathis (1970). *Comptes Rend. Acad. Sci. Paris*, **271D**, 1094–1096.

Breton, J., M. Michel-Villaz and G. Paillotin (1973). *Biochim. Biophys. Acta*, **314**, 42–56.

Brody, S. S. (1975). *Z. Naturfors.*, **30b**, 318—322.

Büchel, K. H. and G. Schäfer (1970). *Z. Naturfors.*, **25b**, 1465–1474.

Bucke, C. and J. Coombs (1974). In M. Avron (Ed.), *Proc. 3rd Int. Congr. Photosyn.*, Elsevier, Amsterdam, pp. 1567–1571.

Calvin, M. and J. A. Bassham (1962). *The Photosynthesis of Carbon Compounds*, Benjamin, New York.

Carrier, J. M. (1967). In T. W. Goodwin (Ed.), *Biochemistry of Chloroplasts*, Vol. 2, Academic Press, London, pp. 551–557.

Chan, P. H. and S. G. Wildman (1972). *Biochim. Biophys. Acta*, **277**, 677–680.

Clayton, R. K. (1965). *Molecular Physics in Photosynthesis*, Blaisdell, New York, p. 39.

Clayton, R. K. (1970). *Light and Living Matter*, Vol. 1, McGraw-Hill, New York.

Clayton, R. K. (1971). *Light and Living Matter*, Vol. 2, McGraw-Hill, New York.

Cole, J. S. and H. Aleem (1973). *Proc. Natl. Acad. Sci. U.S.A.*, **70**, 3571–3575.

Commoner, B., J. J. Heise and J. Townsend (1956). *Proc. Natl. Acad. Sci. U.S.*, **42**, 710–718.

Cox, R. P. and D. S. Bendall (1972). *Biochim. Biophys. Acta*, **283**, 124–135.

Cramer, W. A. and W. L. Butler (1969). *Biochim. Biophys. Acta*, **172**, 503–510.

Criddle, R. S. (1966). In T. W. Goodwin (Ed.), *Biochemistry of Chloroplasts*, Vol. 1, Academic Press, London, pp. 203–231.

Criddle, R. S., B. Dau, G. E. Keinkopf and R. C. Huffaker (1970). *Biochem. Biophys. Res. Commun.*, **41**, 621–627.

Csatorday, K., E. Lehoczki and L. Szalay (1975). *Biochim. Biophys. Acta*, **376**, 268–273.

Danon, A. and W. Stoeckenius (1974). *Proc. Natl. Acad. Sci. U.S.A.*, **71**, 1234–1238.

Davenport, H. E. (1952). *Nature*, **170**, 1112–1114.

Davenport, H. E. (1972). In G. Forti, M. Avron and A. Melandri (Eds.), *Proc. IInd Int. Cong. on Photosynthesis*, Dr. W. Junk, N.V., The Hague, pp. 1593–1601.

Davenport, H. E. and R. Hill (1952). *Proc. Roy. Soc. B*, **139**, 327–345.

Dawson, A. P. and M. J. Selwyn (1974). In A. T. Bull and others (Eds.), *Companion to Biochemistry*, Longman, London, pp. 553–586.

Deamer, D. W. and A. R. Crofts (1967). *J. Cell Biol.*, **33**, 395–410.

Döring, G. (1975). *Biochim. Biophys. Acta*, **376**, 274–284.

Dougherty, R. C., H. H. Strain, W. A. Svec, R. A. Uphaus and J. J. Katz (1970). *J. Amer. Chem. Soc.*, **92**, 2826–2833.

202

Downton, W. J. S. (1975). *Photosynthetica*, **9**, 96–105.

Duysens, L. N. M., J. Amesz and B. M. Kamp (1961). *Nature, London*, **190**, 510–511.

Emerson, R. (1958). *Ann. Rev. Plant Physiol.*, **9**, 1–24.

Emerson, R. and W. Arnold (1932). *J. Gen. Physiol.*, **15**, 391–420 and **16**, 191–205.

Emerson, R. and M. S. Nishimura (1949). In J. Franck and W. E. Loomis (Eds.), *Photosynthesis in Plants*, Iowa State College Press, Ames, Iowa, pp. 219–238.

Erecinska, M., M. Wagner and B. Chance (1973). In D. R. Sanadi (Ed.), *Current Topics in Bioenergetics*, Vol. 5, Academic Press, New York, pp. 267–303.

Esser, A. F. (1974) *Photochem. and Photobiol.*, **20**, 173–181.

Evans, E. H. and A. R. Crofts (1974). *Biochim. Biophys. Acta*, **357**, 74–88.

Evans, M. C. W., B. B. Buchanan and D. I. Arnon (1966). *Proc. Natl. Acad. Sci.*, **55**, 928.

Evans, M. C. W., R. Cammack and S. G. Reeves (1974). In M. Avron (Ed.), *Proc. IIIrd Int. Cong. on Photosynthesis*, Elsevier, Amsterdam, pp. 383–388.

Evans, M. C. W., C. K. Sihra, J. R. Bolton and R. Cammack (1975). *Nature*, **256**, 668–670.

Everson, R. G. (1971). In M. D. Hatch, C. B. Osmond and R. O. Slatyer (Eds.), *Photosynthesis and Photorespiration*, Wiley–Interscience, New York, pp. 275–282.

Faludi-Daniel, A., S. Demeter and A. S. Garay (1973). *Plant Physiol.*, **52**, 54–56.

Fork, D. C. (1963). In *Photosynthetic Mechanisms of Green Plants*, Publ. No. 1145, N.A.S.–N.R.C., Washington, D.C., pp. 352–361.

Förster, T. (1959). *Discussions Faraday Soc.*, **27**, 7–17.

Forti, G. and G. Zanetti (1967). In T. W. Goodwin (Ed.), *Biochemistry of Chloroplasts*, Vol. 2, Academic Press, London, pp. 523–529.

Forti, G., M. L. Bertole and G. Zanetti (1965). *Biochim. Biophys. Acta*, **109**, 33–40.

Forti, G., L. Rosa and F. Garleschi (1972). *FEBS letters*, **27**, 23–26.

Fowler, C. (1974). *Biochim. Biophys. Acta*, **357**, 308–318.

French, C. S. and H. S. Huang (1957). *Carnegie Inst. Washington Year Book*, **56**, 266–268.

French, C. S. and L. Prager (1969). In H. Metzner (Ed.), *Progress in Photosynthesis Research*, Vol. 2, Institute für Chemische Pflanzenphysiologie Tubingen, pp. 555–564.

Frenkel, A. W. (1970). *Biol. Revs.*, **45**, 569–616.

Fujita, Y. and J. Myers (1967). *Arch. Biochem. Biophys.*, **119**, 8–15.

Garewal, H. S. and A. R. Wasserman (1974). *Biochemistry*, **13**, 4063–4071.

Geacintov, N. E., F. Van-Nostrand, J. F. Becker and J. B. Tinkel (1972). *Biochim. Biophys. Acta*, **267**, 65–79.

Gerster, R., B. Dimon and A. Peybernes (1974). In M. Avron (Ed.), *Proc. IIIrd Int. Cong. on Photosynthesis*, Elsevier, Amsterdam, pp. 1589–1600.

Gibbs, M. (1963). In *Photosynthetic Mechanisms of Green Plants*, publication no. 1145, NAS/NRC, Washington, D.C., p. 663–674.

Glasstone, S. and D. Lewis (1960). *Elements of Physical Chemistry*, 2nd ed., Macmillan, London.

Gorman, D. S. and R. P. Levine (1966). *Plant Physiol.*, **41**, 1648–1656.

Govindjee, G. Papageorgeou and E. Rabinowitch (1967). In G. G. Guilbault (Ed.), *Fluorescence*, Dekker, New York, pp. 511–564.

Govindjee, R., W. R. Smith and Govindjee (1974). *Photochem. and Photobiol.*, **20**, 191–199.

Graeber, P. and H. T. Witt (1974). In M. Avron (Ed.), *Proc. IIIrd Int. Cong. on Photosynthesis Res.*, Vol. 2, Elsevier, Amsterdam, pp. 951–956.

Green, D. E. (1974). *Biochim. Biophys. Acta,* **346,** 27–78.

Gregory, R. P. F. (1975). *Biochem. J.,* **148,** 487–497.

Gregory, R. P. F. and S. Raps (1974). *Biochem. J.,* **142,** 193–201.

Gregory, R. P. F., S. Raps and W. F. Bertsch (1971). *Biochim. Biophys. Acta,* **234,** 330–334.

Haehnel, W. (1974). In M. Avron (Ed.), *Proc. IIIrd Int. Cong. Photosyn.* Elsevier, Amsterdam, pp. 557–568.

Hall, D. O. and K. K. Rao (1972). *Photosynthesis,* Arnold, London.

Hatch, M. D. and C. R. Slack (1966). *Biochem. J.,* **101,** 103–111.

Hauska, G., S. Reimer and A. Trebst (1974). *Biochim. Biophys. Acta,* **357,** 1–13.

Heber, U. (1974). In M. Avron (Ed.), *Proc. IIIrd. Int. Cong. on Photosynthesis,* Elsevier, Amsterdam, pp. 1345–1348.

Heber, U. and M. R. Kirk (1974). *Biochim. Biophys. Acta,* **376,** 136–150.

Heizman, P. (1974). In M. Avron (Ed.), *Proc. IIIrd Int. Cong. on Photosynthesis,* Elsevier, Amsterdam, pp. 1745–1754.

Heldt, H. W., R. Fliege, K. Lehner, M. Milovancev and K. Werden (1974). In M. Avron (Ed.), *Proc. IIIrd Int. Cong. on Photosynthesis,* Elsevier, Amsterdam, pp. 1369–1380.

Henninger, M. D. and F. L. Crane (1967). *J. Biol. Chem.,* **242,** 1155–1159.

Hill, R. (1963). In M. Florkin and E. H. Stotz (Eds.), *Comprehensive Biochemistry,* Vol. 9, Elsevier, Amsterdam, pp. 73–97.

Hill, R. (1965). In P. N. Campbell and G. D. Greville (Eds.), *Essays in Biochemistry,* Vol. 1, Academic Press, London, pp. 121–151.

Hill, R. and D. S. Bendall (1966). In T. W. Goodwin (Ed.), *Biochemistry of Chloroplasts,* Vol. 2, Academic Press, London, pp. 559–564.

Hill, R. and F. Bendall (1960a). *Nature,* **187,** 417.

Hill, R. and F. Bendall (1960b). *Nature,* **186,** 136–137.

Hind, G. and A. T. Jagendorf (1963). *Proc. Natl. Acad. Sci. U.S.A.,* **49,** 715–722.

Hind, G. and J. M. Olson (1967). In *Energy Conversion by the Photosynthetic Apparatus,* Brookhaven Symposium No. 19, pp. 188–194.

Hiyama, T. and B. Ke (1972). In G. Forti, M. Avron and A. Melandri (Eds.), *Proc. IInd. Int. Cong. Photosyn. Res.,* Vol. 1, Dr. W. Junk, N.V., The Hague, pp. 491–497.

Hoch, G. and O.v.H. Owens (1963). In *Photosynthetic Mechanisms of Green Plants,* publication no. 1145, NAS/NRC, Washington, D.C., pp. 409–420.

Izawa, S. and N. E. Good (1965). *Biochim. Biophys. Acta,* **102,** 20–38.

Izawa, S. and N. E. Good (1966). *Plant Physiol.,* **41,** 544–552.

Izawa, S. and N. E. Good (1968). *Biochim. Biophys. Acta,* **162,** 380–391.

Izawa, S., R. L. Heath and G. Hind (1969). *Biochim. Biophys. Acta,* **180,** 388–398.

Jackson, J. B. and A. R. Crofts (1969). *FEBS letters,* **4,** 185–189.

Jagendorf, A. T. (1975). In Govindjee (Ed.), *Bioenergetics of Photosynthesis,* Academic Press, New York, pp. 413–492.

Jensen, R. G. and J. T. Bahr (1974). In M. Avron (Ed.), *Proc. IIIrd Int. Cong. on Photosynthesis,* Elsevier, Amsterdam, pp. 1411–1420.

Johnson, H. S. and M. D. Hatch (1970). *Biochem. J.,* **119,** 273–280.

Joliot, P. and B. Kok (1975). In Govindjee (Ed.), *Bioenergetics of Photosynthesis*, Academic Press, New York, pp. 387–412.

Joliot, P., G. Barbieri and R. Chabaud (1969). *Photochem. Photobiol.*, **10**, 309–329.

Kamen, M. D. (1963). *Primary Processes in Photosynthesis*, Academic Press, New York, p. 4.

Katoh, S., I. Shiratori and A. Takamiya (1962). *J. Biochem. (Tokyo)*, **51**, 32–40.

Ke, B. (1973). *Biochim. Biophys. Acta*, **301**, 1–33.

Kirk, J. T. O. (1966). In T. W. Goodwin (Ed.), *Biochemistry of Chloroplasts*, Vol. 1, Academic Press, London, pp. 319–340.

Klingenberg, M. (1968). In T. P. Singer (Ed.), *Biological Oxidations*, Interscience, New York, pp. 3–54.

Klingenberg, M., A. Müller, P. Schmidt-Mende and H. T. Witt (1962). *Nature*, **194**, 379–380.

Knaff, D. B. and D. I. Arnon (1969a). *Proc. Natl. Acad. Sci. U.S.*, **63**, 963–969.

Knaff, D. B. and D. I. Arnon (1969b). *Proc. Natl. Acad. Sci. U.S.*, **64**, 715–722.

Knaff, D. B. and B. B. Buchanan (1975). *Biochim. Biophys. Acta*, **376**, 549–560.

Kok, B. (1951). In *Carbon Dioxide Fixation and Photosynthesis*, Symp. Soc. Exp. Biol. 5, Cambridge University Press, pp. 211–221.

Kok, B. (1961). *Biochim. Biophys. Acta*, **48**, 527–533.

Kok, B. (1966). In J. B. Thomas and J. C. Goedheer (Eds.), *Currents in Photosysnthesis*, Donker, Rotterdam, pp. 383–392.

Kok, B. and G. M. Cheniae (1966). In D. R. Sanadi (Ed.), *Current Topics in Bioenergetics*, Vol. 1, Academic Press, New York, pp. 1–47.

Kok, B. and E. A. Datko (1965). *Plant Physiol*, **40**, 1171–1177.

Kok, B., B. Forbush and M. McGloin (1970). *Photochem. Photobiol.*, **11**, 457–475.

Kortschak, H. P., C. E. Hartt and G. O. Burr (1965). *Plant Physiol.* **40**, 209–213.

Krasnovsky, A. A. (1969). In H. Metzner (Ed.), *Progress in Photosynthesis Research*, Vol. 2, Institut für Chemische Pflanzenphysiologie, Tubingen, pp. 709–727.

Kreutz, W. (1966). In T. W. Goodwin (Ed.), *Biochemistry of Chloroplasts*, Vol. 1, Academic Press, London, pp. 83–88.

Laetsch, W. M. (1974). *Ann. Rev. Plt. Physiol.*, **25**, 27–52.

Lardy, H. A. and S. M. Ferguson (1969). *Ann. Rev. Biochem.*, **38**, 991–1034.

Lavorel, J. (1959). *Plant Physiol.*, **34**, 204–209.

Levine, R. P. (1969). In H. Metzner (Ed.), *Progress in Photosynthesis Research*, Vol. 2, Institut für Chemische Pflanzenphysiologie, Tubingen, pp. 971–977.

Lichtenthaler, H. K. and R. B. Park, (1963). *Nature*, **198**, 1070–1072.

Ljones, T. (1974). In A. Quisnel (Ed.), *The Biology of Nitrogen Fixation*, North-Holland, Amsterdam, pp. 617–638.

McCarty, R. E. and E. Racker (1967). *Brookhaven Symposium No. 19*, pp. 202–214.

McIntosh, A. R., M. Chu and J. R. Bolton (1975). *Biochim. Biophys. Acta*, **376**, 308–314.

MacRobbie, E. A. (1965). *Biochim. Biophys. Acta*, **94**, 64–73.

Mahler, E. H. and H. R. Cordes (1971). *Biological Chemistry*, Harper and Row, New York, pp. 27–32 (1st ed., 1966).

Malkin, R. and A. J. Bearden (1971). *Proc. Natl. Acad. Sci., U.S.*, **68**, 16–19.

Malkin, R. and A. J. Bearden (1975). *Quart. Rev. Biophys.*, **7**, 131–177.

Massey, V., R. G. Matthews, G. P. Foust, L. G. Howell, C. H. Williams, G. Zanetti and S. Ronchi (1970). In H. Sund (Ed.), *Pyridine Nucleotide-dependent Dehydrogenases*, Springer-Verlag, Berlin, pp. 393–411.

Mayne, B. C. (1969). In H. Metzner (Ed.), *Progress in Photosynthesis Research*, Vol. 2, Institut für Chemische Pflanzenphysiologie, Tubingen, pp. 947–951.

Michel, J. M. and M. R. Michel-Wolwertz (1969). In H. Metzner (Ed.), *Progress in Photosynthesis Research*, Vol. 1, Institut für Chemische Pflanzenphysiologie, Tubingen, pp. 115–127.

Mitchell, P. (1961). *Nature*, **191**, 144.

Mitchell, P. (1966). *Biol. Rev.* **41**, 445–502.

Mitchell, R. A., R. D. Hill and P. D. Boyer (1967). *J. Biol. Chem.*, **242**, 1793–1801.

Morita, S. (1968). In K. Shibata and others (Eds.), *Comparative Biochemistry and Biophysics of Photosynthesis*, University of Tokyo Press, Tokyo, pp. 133–139.

Mortenson, L. E., R. E. Valentine and J. E. Carnahan (1962). *Biochem. Biophys. Res. Commun.*, **7**, 448–452.

Mühlethaler, K. (1966). In T. W. Goodwin. (Ed.), *Biochemistry of Chloroplasts*, Vol., 1, Academic Press, London, pp. 49–64.

Murata, N. (1969). *Biochim. Biophys. Acta*, **189**, 171–181.

O'hEocha, C. (1966). In T. W. Goodwin (Ed.), *Biochemistry of Chloroplasts*, Vol. 1, Academic Press, London, pp. 407–421.

Olson, J. M. and E. K. Stanton (1966). In L. P. Vernon and G. R. Seely (Eds.), *The Chlorophylls*, Academic Press, New York, pp. 381–398.

Olson, R. A. (1963). In *Photosynthetic Mechanisms of Green Plants*, Publication no. 1145, NAS/NRC, Washington, D.C., pp. 545–559.

Oswald, W. J. (1973). *Solar Energy*, **15**, 107–117.

Park, R. B. and J. Biggins (1964). *Science*, **144**, 1009–1011.

Park, R. B. and D. Branton (1967). In *Energy Conversion by the Photosynthetic Apparatus*, Brookhaven symposium No. 19, Brookhaven National Laboratory, New York, pp. 341–352.

Park, R. B. and N. G. Pon (1961). *J. Molec. Biol.*, **3**, 1–10.

Parson, W. W. (1968). *Biochim. Biophys. Acta*, **153**, 248–259.

Parson, W. W. and G. D. Case (1970). *Biochim. Biophys. Acta*, **205**, 232–245.

Parson, W. W. and R. J. Cogdell (1975). *Biochim. Biophys. Acta*, **416**, 105–149.

Rabinowitch, E. (1945a). *Photosynthesis and Related Processes*, Vol, 1, Interscience, New York, pp. 111–127.

Rabinowitch, E. (1945b). *Photosynthesis and Related Processes*, Vol, 1, Interscience, New York, pp. 328–330; 531–532.

Rabinowitch, E. (1951). *Photosynthesis and Related Processes*, Vol, 2, Interscience, New York.

Rabinowitch, E. and Weiss, J. (1937). *Proc. Roy. Soc. A*, **162**, 251–267.

Racker, E. and T. E. Conover (1963). *Fed. Proc.*, Pt. 1, **22**, 1088–1091.

Reeves, S. G. and D. O. Hall (1973). *Biochim. Biophys. Acta*, **314**, 66–78.

Reeves, S. G., D. O. Hall and J. West (1972). In G. Forti, M. Avron and A. Melandri (Eds.), *Proc. IInd Int. Congr. Photosynthesis Res.*, Dr. W. Junk N.V., The Hague, pp. 1357–1369.

Renger, G. (1971). *Biochim. Biophys. Acta*, **256**, 428–439.

Ridley, S. M. and R. M. Leech (1970). *Nature,* **227,** 463–465.

Riley, G. A. (1944). *Amer. Scientist,* **32,** 132–134.

Rumberg, B. and U. Siggel (1969). *Naturwissenschaften,* **56,** 130–132.

Ryrie, I. J. and A. T. Jagendorf (1971). *J. Biol. Chem.,* **246,** 3771–3774.

Saha, S., R. Ouitrakul, S. Izawa and N. E. Good (1971). *J. Biol. Chem.,* **246,** 3204–3209.

San Pietro, A. and H. M. Lang (1958). *J. Biol. Chem.,* **231,** 211–229.

Schiff, J. A. (1973). *Advan. Morphogenesis,* **10,** 265–312.

Schiff, J. A. (1974). In M. Avron (Ed.), *Proc. IIIrd Int. Cong. on Photosynthesis,* Elsevier, Amsterdam, pp. 1691–1717.

Schmid, G. H. and H. Gaffron (1968). *J. Gen. Physiol.,* **52,** 212.

Scott, B. (1974). Ph.D. thesis, University of Manchester, pp. 122ff.

Seliger, H. H. and W. D. McElroy (1965). *Light, Physical and Biological Action,* Academic Press, New York, pp. 79–118.

Shahak, Y., H. Hardt and M. Avron (1975). *FEBS letters,* **54,** 151–154.

Shin, M., K. Tagawa and D. I. Arnon (1963). *Biochem. Z.,* **338,** 84–96.

Sinclair, J. (1972). *Plant Physiol.,* **50,** 778–783.

Sironval, C., H. Clijsters, J-M. Michel, R. Bronchart and R. M. Michel-Wolwertz (1966). In J. B. Thomas and J. C. Goedheer (Eds.), *Currents in Photosynthesis,* Donker, Rotterdam, pp. 111–120.

Slooten, L. (1972). *Biochim. Biophys. Acta,* **275,** 208–218.

Smith, B. N. and M. J. Robbins (1974). In M. Avron (Ed.), *Proc. IIIrd Int. Cong. on Photosynthesis,* Elsevier, Amsterdam, pp. 1579–1587.

Smith, J. H. and A. Benitez (1955). *Modern Methods of Plant Analysis,* Vol. 4, Springer, Berlin, p. 142.

Smith, L. (1968). In. T. P. Singer (Ed.), *Biological Oxidations,* Interscience, New York, pp. 55–122.

Staehelin, L. A., P. A. Armond and K. R. Miller (1976). *Brookhaven Symp.,* in press.

Stiehl, H. H. and H. T. Witt (1968). *Z. Naturforschung,* **23b,** 220–224.

Stokes, D. M. and D. A. Walker (1971). In M. D. Hatch, C. B. Osmond and R. O. Slatyer (Eds.), *Photosynthesis and Photorespiration,* Wiley–Interscience, New York, pp. 226–231.

Strain, H. H. (1966). In T. W. Goodwin (Ed.), *Biochemistry of Chloroplasts,* Vol. I, Academic Press, London, pp. 387–406.

Strehler, B. and W. Arnold (1951). *J. Gen. Physiol.,* **34,** 809–820.

Strittmatter, P. (1968). In T. P. Slater (Ed.), *Biological Oxidations,* Interscience, New York, pp. 171–191.

Stuart, A. L. and A. R. Wasserman (1975). *Biochim. Biophys. Acta,* **376,** 561–572.

Sybesma, C. and B. Kok (1969). *Biochim. Biophys. Acta,* **180,** 410–413.

Tanada, T. (1951). *Amer. J. Bot.,* **38,** 276–283.

Tanner, W. and O. Kandler (1969). In H. Metzner (Ed.), *Progress in Photosynthesis Research,* Vol. 3, Institut für Chemische Pflanzenphysiologie, Tubingen, pp. 1217–1223.

Tanner, W., E. Loos and O. Kandler (1966). In J. B. Thomas and J. C. Goedheer (Eds.), *Currents in Photosynthesis,* Donker, Rotterdam, pp. 243–251.

Taylor, D. L. (1971). *Comp. Biochem. Physiol.,* **38A,** 233–236.

Thomas, J. B. (1965). *Primary Photoprocesses in Biology*, North-Holland, Amsterdam, Chapters 1 and 2.

Thomas, J. B., J. H. van Lierop and M. Ten Ham (1966). *Biochim. Biophys. Acta*, **143**, 204–220.

Thornber, J. P. and H. R. Highkin (1974). *Eur. J. Biochem.*, **41**, 109–116.

Thornber, J. P., R. P. F. Gregory, C. A. Smith and J. L. Bailey (1967a). *Biochemistry*, **6**, 391–396.

Thornber, J. P., J. C. Stewart, M. C. W. Hatton and J. L. Bailey (1967b). *Biochemistry*, **6**, 2006–2014.

Tien, H. T. (1974). *Bilayer Lipid Membranes (BLM): Theory and Practice*, Dekker, New York, pp. 348–377.

Tolbert, N. E. (1971). *Ann. Rev. Plant Physiol.*, **22**, 45–74.

Tolbert, N. E. and F. J. Ryan (1974). In M. Avron (Ed.), *Proc. IIIrd Int. Cong. on Photosynthesis*, Elsevier, Amsterdam, pp. 1303–1309.

Trebst, A. (1974). *Ann. Rev. Plant Physiol.*, **25**, 423–458.

Treharne, R. W., T. E. Brown and L. P. Vernon (1963). *Biochim. Biophys. Acta*, **75**, 324–332.

Van Niel, C. B. (1935). *Cold Spring Harbour Symp. Quant. Biol.*, **3**, 138–149.

Vernon, L. P., E. R. Shaw and B. Ke (1966). *J. Biol. Chem.*, **241**, 4101–4109.

Walker, D. A. (1974). In D. H. Northcote (Ed.), M.T.P. Int. Rev. of Science, Biochemistry Ser. I, Vol. 11: *Plant Biochemistry*, Butterworth, London, pp. 1–49.

Warburg, O. (1920). *Biochem. Zeits.*, **103**, 188–217.

Warburg, O. and W. Lüttgens (1944). *Naturwissenschaften*, **32**, 301.

Warburg, O., G. Krippahl and A. Lehman (1969). *Amer. J. Bot.*, **56**, 961–971.

Warden, J. T. and J. R. Bolton (1974). *Photochem. and Photobiol.*, **20**, 251–262.

Weaver, E. C. and H. E. Weaver (1972). In A. C. Giese (Ed.), *Photophysiology*, Vol. 8, Academic Press, New York, pp. 1–32.

Weaver, P., K. Tinker and R. C. Valentine (1965). *Biochem. Biophys. Res. Commun.*, **21**, 195–201.

Weier, T. E. and A. A. Benson (1966). In T. W. Goodwin (Ed.), *Biochemistry of Chloroplasts*, Vol. I, Academic Press, London, pp. 91–113.

Werden, K., H. W. Heldt and M. Milovancev (1975). *Biochim. Biophys. Acta*, **396**, 276–292.

Wessels, J. S. C. and G. Voorn (1972). In G. Forti, M. Avron and A. Melandri (Eds.), *Proc. IInd Int. Cong. Photosyn. Res.*, Dr. W. Junk, N.V. The Hague, pp. 833–845.

Wilson, D. F., P. L. Dutton and M. Wagner (1973). In D. R. Sanadi (Ed.), *Current Topics in Bioenergetics*, Vol. 5, Academic Press, New York, pp. 233–265.

Witt, H. T. (1971). *Quart. Rev. Biophysics*, **4**, 365–477.

Witt, H. T. (1975). In Govindjee (Ed.), Bioenergetics of Photosynthesis, Academic, New York, 493–554.

Witt, H. T., B. Suerra and J. Vater (1966). In J. B. Thomas and J. C. Goedheer (Eds.), *Currents in Photosynthesis*, Donker, Rotterdam, pp. 272–283.

Witt, H. T., B. Rumberg and W. Junge (1969a). In H. Staudinger and B. Hess (Eds.), *Biochemie des Sauerstoffs*, Mosbach Symposium 1968, Springer, Berlin, pp. 262–306.

Witt, H. T., B. Rumberg, W. Junge, G. Döring, H. H. Stiehl, J. Weikard and C. Wolff (1969b). In H. Metzner (Ed.), *Progress in Photosynthesis Research*, Vol. 3, Institut für

Chemische Pflanzenphysiologie, Tubingen, pp. 1361–1373.

Wolff, C., M. Gläser and H. T. Witt (1974). In M. Avron (Ed.), *Proc. IIIrd Int. Cong. Photosyn.*, Elsevier, Amsterdam, pp. 295–305.

Woo, K. C., J. M. Anderson, N. K. Boardman, W. J. S. Downton, C. B. Osmond and S. W. Thorne (1970). *Proc. Natl. Acad. Sci. U.S.*, **67**, 18–25.

Wood, P. M. and D. S. Bendall (1975). *Biochim. Biophys. Acta,* **387**, 115–128.

Wraight, C. A., R. J. Cogdell and R. K. Clayton (1975). *Biochim. Biophys. Acta*, **396**, 242–249.

Yoch, D. C. and D. I. Arnon (1974). In A. Quisnel (Ed.), *The Biology of Nitrogen Fixation*, North-Holland, Amsterdam, pp. 687–695.

Young, J. H., E. F. Korman and J. McLick (1974). *Bio-org. Chem.*, **3**, 1–15.

Zelitch, I. (1971). *Photosynthesis, Photorespiration and Plant Productivity*, Academic Press, New York, pp. 173–212.

Zweig, G. and M. Avron (1965). *Biochem. Biophys. Res. Commun.*, **19**, 397–400.

Author Index

Subject Index

214